Fun with

ROMAN NUMERALS

Math Workbook

$$\sqrt{C} = X$$

$$XV \div V = III$$

$$II^{IV} = XVI$$

Chris McMullen, Ph.D.

Fun with Roman Numerals Math Workbook

Chris McMullen, Ph.D.

Zishka Publishing

ISBN: IX-VII-VIII-I-IX-IV-I-VI-IX-I-V-VII-I

Mathematics > Roman Numerals

Education > Math > Arithmetic

CONTENTS

INTRODUCTION

The first three chapters of this book are aimed at building fluency with reading and writing Roman numerals, so that the student can quickly understand a numeral like XLIV or quickly write a number like 449 as a Roman numeral. Once the student has become fluent in Roman numerals, he or she is ready to enjoy the pattern puzzles, variety puzzles, or arithmetic problems included in the remaining chapters. The answer to every problem or puzzle can be found in the Answer Key at the back of the book. May you become more fluent with Roman numerals and enjoy the exercises in this book. Let the games begin! (Colosseum, togas, gladiators, and lions not included, but feel free to take a field trip to Italy.)

I

Converting to Roman Numerals

Chapter goals:

- Learn how Roman numerals work.
- Be able to convert any whole number up to 3999 into a Roman numeral.
- Practice writing Roman numerals to build fluency.

How Roman Numerals Work

First memorize the values of the following letters. (Otherwise, Roman numerals will be totally Greek to you.)

$$I = 1 \quad, \quad V = 5 \quad, \quad X = 10 \quad, \quad L = 50$$
$$C = 100 \quad, \quad D = 500 \quad, \quad M = 1000$$

The letters I, X, C, and M may appear up to three times in a row in order to make up to three times their value, like the examples below. However, don't repeat any numerals that include fives, such as V, L, or D.

$$II = 2 \quad, \quad III = 3 \quad, \quad XX = 20 \quad, \quad XXX = 30$$
$$CC = 200 \quad, \quad CCC = 300 \quad, \quad MM = 2000 \quad, \quad MMM = 3000$$

The numbers 4, 9, 40, 90, 400, and 900 are made as follows. In these cases, the smaller numeral on the left is subtracted from the larger numeral on the right. Don't make any other numbers by subtracting. These are special.

$$IV = 4 \quad, \quad IX = 9 \quad, \quad XL = 40 \quad, \quad XC = 90$$
$$CD = 400 \quad, \quad CM = 900$$

As we will explore throughout this chapter, other numbers are made by adding smaller Roman numerals to the right of larger Roman numerals. For example, CV = 105.

Less Than Forty

To write a Roman numeral that is less than forty, first ask how many tens are needed. For example, 27 needs two tens, 33 needs three tens, 19 needs one ten, and 7 doesn't need a ten. Write an X for each ten. Now write a number from one to nine to the right of the X's. Recall that IV = 4 and IX = 9. Note that VI = 5 + 1 = 6, VII = 5 + 2 = 7, and VIII = 5 + 3 = 8.

Examples. Convert each number to a Roman numeral.

- 32 = 30 + 2 = XXX + II = XXXII.
- 26 = 20 + 6 = XX + VI = XXVI.
- 8 = 5 + 3 = V + III = VIII.
- 14 = 10 + 4 = X + IV = XIV.

Problems. Convert each number to a Roman numeral.

(a) 30 = XXX

(b) 15 = XV

(c) 3 = III

(d) 21 = XXI

(e) 19 = XXXXXIX

(f) 37 = XXXVII

(g) 28 = XXVIII

(h) 12 = XII

(i) 36 = XXXVI

(j) 24 = XXIV

Forty to One Hundred

Note that L = 50, XL = 40, and XC = 90. To make numbers in the sixties, seventies, or eighties, add one to three X's to the right of an L. That is, LX = 60, LXX = 70, and LXXX = 80. As in the previous section, now write a number from one to nine to the right of the numeral for the tens.

Examples. Convert each number to a Roman numeral.

- 51 = 50 + 1 = L + I = LI.
- 77 = 50 + 20 + 7 = L + XX + VII = LXXVII.
- 48 = 40 + 8 = XL + VIII = XLVIII.
- 84 = 50 + 30 + 4 = L + XXX + IV = LXXXIV.
- 93 = 90 + 3 = XC + III = XCIII.

Problems. Convert each number to a Roman numeral.

(a) 72 = LXXII

(b) 58 = LVIII

(c) 67 = LXVII

(d) 45 = XLV

(e) 91 = ~~LXX~~ XCI

(f) 83 = LXXXIII

(g) 49 = XLIX

(h) 64 = LXIV

(i) 86 = LXXX

(j) 99 = XCIX

One Hundred to Four Hundred

To make a Roman numeral between one hundred and four hundred, first write one C for each hundred. For example, 235 needs two C's while 376 needs three C's. After writing the C's, write a number from one to ninety-nine to the right of the C's.

Examples. Convert each number to a Roman numeral.

- $124 = 100 + 20 + 4 = C + XX + IV = CXXIV.$
- $205 = 200 + 5 = CC + V = CCV.$
- $383 = 300 + 80 + 3 = CCC + LXXX + III = CCCLXXXIII.$
- $191 = 100 + 90 + 1 = C + XC + I = CXCI.$
- $252 = 200 + 50 + 2 = CC + L + II = CCLII.$

Problems. Convert each number to a Roman numeral.

(a) 110 = CX

(b) 328 = CCCXXVIII

(c) 222 = CCXXII

(d) 286 = CCLXXXVI

(e) 136 = CXXXVI

(f) 300 = CCC

(g) 375 = CCCLXXV

(h) 194 = CXCIV

(i) 150 = CL

(j) 349 = CCCXLIX

Four Hundred to One Thousand

Note that D = 500, CD = 400, and CM = 900. To make six hundred, seven hundred, or eight hundred, add one to three C's to the right of a D. That is, DC = 600, DCC = 700, and DCCC = 800. Now write a number from one to ninety-nine to the right of the numeral for the hundreds.

Examples. Convert each number to a Roman numeral.

- 526 = 500 + 20 + 6 = D + XX + VI = DXXVI.
- 741 = 700 + 40 + 1 = DCC + XL + I = DCCXLI.
- 459 = 400 + 50 + 9 = CD + L + IX = CDLIX.
- 673 = 600 + 70 + 3 = DC + LXX + III = DCLXXIII.
- 905 = 900 + 5 = CM + V = CMV.

Problems. Convert each number to a Roman numeral.

(a) 638 =   (b) 810 =

(c) 984 =   (d) 492 =

(e) 549 =   (f) 517 =

(g) 839 =   (h) 464 =

(i) 777 =   (j) 999 =

One Thousand to Four Thousand

To make a Roman numeral between one thousand and four thousand, first write one M for each thousand. For example, 2548 needs two M's while 3149 needs three M's. After writing the M's, write a number from one to nine hundred ninety-nine to the right of the M's.

Examples. Convert each number to a Roman numeral.

- $1352 = 1000 + 352 = M + CCC + LII = MCCCLII.$
- $2718 = 2000 + 718 = MM + DCCXVIII = MMDCCXVIII.$
- $3056 = 3000 + 56 = MMM + LVI = MMMLVI.$
- $2143 = 2000 + 143 = MM + CXLIII = MMCXLIII.$
- $1637 = 1000 + 637 = M + DCXXXVII = MDCXXXVII.$

Problems. Convert each number to a Roman numeral.

(a) $1714 =$ (b) $2625 =$

(c) $3050 =$ (d) $1111 =$

(e) $2027 =$ (f) $3289 =$

(g) $3921 =$ (h) $2343 =$

(i) $1568 =$ (j) $3498 =$

Additional Practice

Test your fluency with Roman numerals.

Problems. Convert each number to a Roman numeral.

(a) 35 = XXXV

(b) 29 = XXIX

(c) 44 = XLIV

(d) 61 = LX

(e) 88 = LXXVIII

(f) 90 = XC

(g) 97 = XCVII

(h) 113 = CXIII

(i) 126 = CXXVI

(j) 172 = CLXXII

(k) 201 = CCI

(l) 218 = CCXVIII

(m) 350 = CCCXXX

(n) 387 = CCCLXXXVII

(o) 484 = CDLXXXIV

(p) 555 = DLV

(q) 719 = DCCXIX

(r) 926 = CMXXVI

(s) 1023 = MXXIII

(t) 1242 = MCCXLII

(u) 1897 = MDCCCXC

(v) 1974 = MCMLXX

(w) 2456 = MMCDLVI

(x) 2735 = MM DCCXXXV

(y) 2999 = MMCMXCIX

(z) 3418 = MMMCDXVIII

Irregular Additions

Roman numerals with four consecutive symbols (like IIII) are considered irregular, but have been used. The modern convention is to use IV, IX, XL, XC, CD, and CM instead.

Examples. Convert each number to Roman numerals using both irregular and regular additions.

- 40 = "XXXX" (irregular) and XL (regular).
- 14 = "XIIII" (irregular) and XIV (regular).
- 900 = "DCCCC" (irregular) and CM (regular).

Problems. Convert each number to Roman numerals using both irregular and regular additions.

(a) 9 =   and

(b) 400 =   and

(c) 90 =   and

(d) 1042 =   and

(e) 29 =   and

(f) 494 =   and

(g) 3900 =   and

Irregular Subtractions

Normal subtractions include IV, IX, XL, XC, CD, and CM. It is sometimes tempting to make Roman numerals with other kinds of subtractions, like "VC," but those are irregular.

Examples. Convert each number to Roman numerals using both irregular and regular subtractions.

- $95 = 100 - 5 =$ "VC" should be $90 + 5 =$ XCV.
- $199 = 100 - 1 =$ "CIC" should be $100 + 90 + 9 =$ CXCIX.
- $950 = 1000 - 50 =$ "LM" should be $900 + 50 =$ CML.

Problems. Convert each number to Roman numerals using both irregular and regular subtractions.

(a) $495 =$ should be

(b) $549 =$ should be

(c) $1950 =$ should be

(d) $145 =$ should be

(e) $450 =$ should be

(f) $990 =$ should be

(g) $499 =$ should be

II

Converting from Roman Numerals

Chapter goals:

- Learn how to read Roman numerals.
- Be able to convert any Roman numeral up to MMMCMXCIX into a familiar whole number.
- Practice converting Roman numerals into whole numbers to build fluency.

Reading Roman Numerals

First review these basic Roman numerals. It is essential to memorize these values.

$$I = 1 \quad , \quad V = 5 \quad , \quad X = 10 \quad , \quad L = 50$$
$$C = 100 \quad , \quad D = 500 \quad , \quad M = 1000$$

When a Roman numeral appears two or more times in a row, the values are added together, like the examples below.

$$II = 2 \quad , \quad III = 3 \quad , \quad XX = 20 \quad , \quad XXX = 30$$
$$CC = 200 \quad , \quad CCC = 300 \quad , \quad MM = 2000 \quad , \quad MMM = 3000$$

If a smaller Roman numeral appears to the left of a larger Roman numeral, it is subtracted from the Roman numeral to its right.

$$IV = 4 \quad , \quad IX = 9 \quad , \quad XL = 40 \quad , \quad XC = 90$$
$$CD = 400 \quad , \quad CM = 900$$

When a smaller Roman numeral appears to the right of a larger Roman numeral, add their values together.

$$CLXVII = 100 + 50 + 10 + 5 + 1 + 1 = 167$$

Note that the subtraction and addition rules may both be involved, like the example below.

$$CXLIX = 100 + (50 - 10) + (10 - 1) = 100 + 40 + 9 = 149$$

Less Than Forty

A Roman numeral that doesn't have an L, C, D, or M has a value that is less than forty. Begin by looking for possible subtractions. For a value under forty, these are IV = 4 and IX = 9. After doing the subtractions, add the remaining Roman numerals, like the examples below.

Examples. Convert each Roman numeral.

- XVI = 10 + 5 + 1 = 16.
- XXXIV = XXX + 4 = 30 + 4 = 34.
- VII = 5 + 2 = 7.
- XXIX = XX + 9 = 20 + 9 = 29.
- XIII = 10 + 3 = 13.

Problems. Convert each Roman numeral.

(a) XXIII =   (b) VI =

(c) XVII =   (d) XIV =

(e) XXV =   (f) XXXI =

(g) XXXVIII =   (h) XX =

(i) XXIV =   (j) XXXIX =

Forty to One Hundred

Recall that L = 50. First check to see if the Roman numeral includes any subtractions. For a value under one hundred, these are IV = 4, IX = 9, XL = 40, and XC = 90. After doing the subtractions, add the remaining Roman numerals, like the examples below.

Examples. Convert each Roman numeral.

- LVII = 50 + 5 + 2 = 57.
- XLVI = 40 + VI = 40 + 6 = 46.
- LXXIV = LXX + 4 = 70 + 4 = 74.
- XCII = 90 + II = 90 + 2 = 92.
- LXXXVIII = 50 + 30 + 5 + 3 = 88.

Problems. Convert each Roman numeral.

(a) LXVIII =   (b) LV =

(c) XLVII =   (d) LXXIII =

(e) XC =   (f) XLIV =

(g) LIII =   (h) LXXXVII =

(i) LXXVI =   (j) XCIV =

One Hundred to Four Hundred

Recall that C = 100. First check to see if the Roman numeral includes any subtractions. For a value under four hundred, these are IV = 4, IX = 9, XL = 40, and XC = 90. After doing the subtractions, add the remaining Roman numerals, like the examples below.

Examples. Convert each Roman numeral.

- CLXVII = 100 + 50 + 10 + 5 + 2 = 167.
- CCXIX = CCX + 9 = 200 + 10 + 9 = 219.
- CCCXCIII = CCC + 90 + III = 300 + 90 + 3 = 393.
- CCLXXXVII = 200 + 50 + 30 + 5 + 2 = 287.
- CXLI = C + 40 + I = 100 + 40 + 1 = 141.

Problems. Convert each Roman numeral.

(a) CLXVI =

(b) CCXXXVIII =

(c) CXCIII =

(d) CCCLXV =

(e) CCLIV =

(f) CXXIX =

(g) CCXLIX =

(h) CCLXXXII =

(i) CXVI =

(j) CCCXCIV =

Four Hundred to One Thousand

Recall that $D = 500$. First check to see if the Roman numeral includes any subtractions. For a value under one thousand, these are $IV = 4$, $IX = 9$, $XL = 40$, $XC = 90$, $CD = 400$, and $CM = 900$. After doing the subtractions, add the remaining Roman numerals, like the examples below.

Examples. Convert each Roman numeral.

- $DXVI = 500 + 10 + 5 + 1 = 516$.
- $DCCCLIX = DCCCL + 9 = 500 + 300 + 50 + 9 = 859$.
- $CDLXXII = 400 + LXXII = 400 + 50 + 20 + 2 = 472$.
- $DCXIII = 500 + 100 + 10 + 3 = 613$.
- $CMLVI = 900 + LVI = 900 + 50 + 6 = 956$.

Problems. Convert each Roman numeral.

(a) DCLXI =

(b) DCCCXXXIII =

(c) CDLXII =

(d) CMXCI =

(e) DCCXIV =

(f) DLXXVII =

(g) CMXXIX =

(h) DCCLX =

(i) DV =

(j) CDXLIV =

One Thousand to Four Thousand

Recall that M = 1000. First check to see if the Roman numeral includes any of the following subtractions: IV = 4, IX = 9, XL = 40, XC = 90, CD = 400, and CM = 900. After doing the subtractions, add the remaining Roman numerals, like the examples below.

Examples. Convert each Roman numeral.

- MDCCLXII = 1000 + 500 + 200 + 50 + 10 + 2 = 1762.
- MMCMXXI = MM + 900 + XXI = 2000 + 900 + 21 = 2921.
- MMMLIV = MMML + 4 = 3000 + 50 + 4 = 3054.
- MMCCXXII = 2000 + 200 + 20 + 2 = 2222.
- MCDLXIII = M + 400 + LXIII = 1000 + 400 + 63 = 1463.

Problems. Convert each Roman numeral.

(a) MDCLXVI =

(b) MMMCCIII =

(c) MMMCDLV =

(d) MMCMXXII =

(e) MMCLXXII =

(f) MMMDCCCXLI =

(g) MCMXIX =

(h) MCCCLXXXIII =

(i) MMDXLVII =

(j) MMMCMXCIX =

Additional Practice

Test your fluency with Roman numerals.

Problems. Convert each Roman numeral.

(a) VIII =   (b) IX =

(c) XXVI =   (d) XLII =

(e) LXVI =   (f) LXXXV =

(g) XC =   (h) CIV =

(i) CXL =   (j) CCXXXII =

(k) CCCXCI =   (l) CDXLIX =

(m) DLIII =   (n) DCVI =

(o) DCCL =   (p) DCCCXCVII =

(q) CMXIV =   (r) MXI =

(s) MCCXLII =   (t) MCDXV =

(u) MMXX =   (v) MMDCCL =

(w) MMCMVI =   (x) MMMCLXVI =

(y) MMMCDXL =   (z) MMMCMLIX =

Irregular Additions

Roman numerals that repeat four or more numerals (like IIII) or which repeat numerals that involve fives, fifties, or five hundreds (like VV, LL, or DD) are considered irregular.

Examples. Convert each irregular Roman numeral. Then rewrite the Roman numeral in regular form.

- "IIII" $= 1 + 1 + 1 + 1 = 4$ should be $5 - 1 = $ IV.

- "VV" $= 5 + 5 = 10$ should be $10 = $ X.

- "LLL" $= 50 + 50 + 50 = 150$ should be $100 + 50 = $ CL.

Problems. Convert each irregular Roman numeral. Then rewrite the Roman numeral in regular form.

(a) "VVV" =   should be

(b) "LXXXX" =   should be

(c) "DD" =   should be

(d) "CCCCC" =   should be

(e) "CLLIIII" =   should be

(f) "DCCCCV" =   should be

(g) "XXVVII" =   should be

Irregular Subtractions

Normal subtractions include IV, IX, XL, XC, CD, and CM. It is sometimes tempting to make Roman numerals with other kinds of subtractions, like "IL," but those are irregular.

Examples. Convert each irregular Roman numeral. Then rewrite the Roman numeral in regular form.

- "IL" $= 50 - 1 = 49$ should be $40 + 9 =$ XL $+$ IX $=$ XLIX.
- "LVX" $= 50 + (10 - 5) = 55$ should be $50 + 5 =$ LV.
- "DMV" $= (1000 - 500) + 5 = 505$ should be $500 + 5 =$ DV.

Problems. Convert each irregular Roman numeral. Then rewrite the Roman numeral in regular form.

    (a) "VL" =   should be

    (b) "IC" =   should be

    (c) "XD" =   should be

    (d) "CIL" =   should be

    (e) "MID" =   should be

    (f) "DLC" =   should be

    (g) "MLD" =   should be

III

Reading and Writing Roman Numerals

Chapter goals:

- Practice reading and writing Roman numerals to build fluency.
- Apply Roman numerals to a variety of numbers, such as historical dates.

Roman Numerals in Use

Example. Convert the Roman numeral.

- Pope John XXIII (former head of the Catholic Church).
- Answer: XXIII = 20 + 3 = 23.

Problems. Convert each Roman numeral.

(a) Titan IV (United States rocket)

(b) Louis XIV (former king of France)

(c) Games of the XXVI Olympiad (Olympic games)

(d) Henry VII (former king of England)

(e) on page xxxiv (an introductory page in a book)

(f) the hour hand points to XI (the time on a clock)

(g) Super Bowl XLIX (American football game)

Years

Example. Rewrite the year using Roman numerals.

- Leonardo da Vinci was born: 1452 AD.

- Answer: $1452 = 1000 + 400 + 50 + 2 =$ MCDLII AD.

Problems. Rewrite each year using Roman numerals.

(a) the founding of Rome: 753 BC

(b) completion of the Colosseum: 80 AD

(c) Rome became a republic: 509 BC

(d) Leaning Tower of Pisa was completed: 1372 AD

(e) the decline of Rome began: 410 AD (arguable)

(f) Sistine Chapel was consecrated: 1483 AD

(g) Galileo was born: 1564 AD

Lifetimes

Example. Convert the years.

- Cicero (statesman): CVI BC to XLIII BC.

- CVI $= 100 + 5 + 1 = 106$ and XLIII $= (50 - 10) + 3 = 43$.

- Final answer: 106 BC to 43 BC.

Problems. Convert each year.

(a) Augustus (emperor): LXIII BC to XIV AD

(b) Spartacus (gladiator): CXI BC to LXXI BC

(c) Constantine (emperor): CCLXXII AD to CCCXXXVII AD

(d) Pompey (general): CVI BC to XLVIII BC

(e) Alexander (king): CCCLVI BC to CCCXXIII BC

(f) Genghis Khan (ruler): MCLXII AD to MCCXXVII AD

U.S. Holidays

Example. Rewrite the date using Roman numerals.

- Holiday: Leap year of 2020 is February 29, 2020.
- $29 = 20 + 9 =$ XXIX and $2020 = 2000 + 20 =$ MMXX.
- Final answer: February XXIX, MMXX.

Problems. Rewrite each date using Roman numerals.

(a) Independence Day, July 4, 1777 (a year after 1776)

(b) The first Memorial Day was May 30, 1868

(c) Veterans Day originated on November 11, 1919

(d) New Year's Eve in 1999 was December 31, 1999

(e) Christmas of 1849 was December 25, 1849

(f) Halloween of 2094 will be October 31, 2094

Historical Dates

Example. Convert the historical date.

- Pearl Harbor: December VII, MCMXLI.

- VII = 5 + 2 = 7 and MCMXLI = 1941.

- Final answer: December 7, 1941.

Problems. Convert each historical date.

(a) MLK "I Have a Dream": August XXVIII, MCMLXIII

(b) French Revolution ended: November IX, MDCCXCIX

(c) Coronation of Napoleon I: December II, MDCCCIV

(d) Magna Carta: June XV, MCCXV

(e) Coronation of Charlemagne: December XXV, DCCC

(f) Boston Tea Party: December XVI, MDCCLXXIII

IV
Arithmetic with Roman Numerals

Chapter goals:

- Add, subtract, multiply, and divide numbers written as Roman numerals.

- Do arithmetic with Roman numerals greater than X.

- Perform long division with remainders.

Addition and Subtraction Facts

Example. VIII + III = XI (since 8 + 3 = 11).

Problems. Add or subtract the Roman numerals.

(a) II + II =

(b) VII − V =

(c) VIII − III =

(d) IV + I =

(e) VI − IV =

(f) VI + V =

(g) VII + II =

(h) V − I =

(i) X − II =

(j) VIII + VI =

(k) V + V =

(l) XVI − VII =

(m) III + III =

(n) X − III =

(o) XI − IX =

(p) V + IV =

(q) IX + VIII =

(r) XV − VII =

(s) X − I =

(t) IX + VI =

(u) XI − IV =

(v) VI + III =

(w) VIII + VII =

(x) IX − IV =

(y) IX + V =

(z) XIV − V =

Adding Larger Roman Numerals

Roman numerals already involve addition. For example, LXXIV = L + X + X + (V − I) = 74 and XVII = X + V + II = 17. To add two Roman numerals together, all you need to do is regroup the additions, like the examples below.

Examples. Add the Roman numerals.

- XVII + XVI = XX + V + V + III = XX + X + III = XXXIII.
  Check: 17 + 16 = 33. Note that V + V = X.

- XL + XXV = (L − X) + XX + V = L − X + XX + V
  = L + X + V = LXV. Check: 40 + 25 = 65.

Problems. Add the Roman numerals.

(a) XVI + XII =

(b) XXXVIII + XXI =

(c) XXXV + XV =

(d) LXIV + LXVI =

(e) CLXVI + LXV =

(f) XCIV + XLIII =

(g) CMLXXXIII + CDLXXII =

Subtracting Larger Roman Numerals

To subtract Roman numerals, distribute the minus sign. For example, $XX - (X + I) = X + X - X - I = X - I = IX$. This is the same as $20 - (10 + 1) = 20 - 10 - 1 = 10 - 1 = 9$.

Examples. Subtract the Roman numerals.

- $XXX - XV = XXX - X - V = XX - V = X + X - V = X + V$ $= XV$. Check: $30 - 15 = 15$. Note that $X - V = V$.

- $LX - XL = L + X - (L - X) = L + X - L - (-X) = X + X$ $= XX$. Check: $60 - 40 = 20$. Note that $-(-10) = +10$.

- $L - XII = XXX + XX - X - II = XXX + X - II = XXX + VIII$ $= XXXVIII$. Check: $50 - 12 = 38$. Note that $L = XXX + XX$.

Problems. Subtract the Roman numerals.

(a) $XXXVIII - XXVI =$

(b) $XXXIV - XV =$

(c) $LXXVIII - LXI =$

(d) $LXXII - XXVI =$

(e) $C - LXV =$

(f) $CMXL - CDV =$

Multiplication Facts

Example. VII × V = XXXV (since 7 × 5 = 35).

Problems. Multiply the Roman numerals.

(a) II × III =

(b) II × II =

(c) IV × IV =

(d) III × III =

(e) VI × IV =

(f) V × II =

(g) VII × III =

(h) VIII × IV =

(i) I × II =

(j) VIII × VI =

(k) V × V =

(l) IX × II =

(m) VIII × VIII =

(n) VII × II =

(o) VI × V =

(p) IX × IV =

(q) IX × VIII =

(r) VII × VII =

(s) VI × VI =

(t) IX × VII =

(u) VII × VI =

(v) VIII × III =

(w) IX × VI =

(x) VII × IV =

(y) VIII × VII =

(z) IX × IX =

Division Facts

Example. XXX ÷ V = VI (since 30 ÷ 5 = 6).

Problems. Divide the Roman numerals.

(a) V ÷ V =

(b) XV ÷ V =

(c) XII ÷ III =

(d) XL ÷ VIII =

(e) XLIX ÷ VII =

(f) XXXII ÷ VIII =

(g) XXXVI ÷ IX =

(h) XXVII ÷ IX =

(i) VIII ÷ II =

(j) LXIII ÷ VII =

(k) XLVIII ÷ VI =

(l) LXIV ÷ VIII =

(m) LVI ÷ VII =

(n) LIV ÷ VI =

(o) XVI ÷ IV =

(p) XXXVI ÷ VI =

(q) LXXII ÷ VIII =

(r) LXIII ÷ IX =

(s) XLII ÷ VI =

(t) XXV ÷ V =

(u) LIV ÷ IX =

(v) LVI ÷ VIII =

(w) XX ÷ IV =

(x) XLII ÷ VII =

(y) XLVIII ÷ VIII =

(z) LXXII ÷ IX =

Multiplying Larger Roman Numerals

Apply the distributive property, $a \times (b + c) = a \times b + a \times c$, and the f.o.i.l. method, $(a + b) \times (c + d) = a \times c + a \times d + b \times c + b \times d$. (f.o.i.l. stands for first, outside, inside, last.)

Examples. Multiply the Roman numerals.

- XVIII × VI = (X + VIII) × VI = X × VI + VIII × VI
  = LX + XLVIII = L + X + (L − X) + VIII = L + L + VIII
  = CVIII. Check: $18 \times 6 = (10 + 8) \times 6 = 60 + 48 = 108$.

- XV × XII = (X + V) × (X + II) = X × X + X × II + V × X + V × II
  = C + XX + L + X = CLXXX. Check: $15 \times 12 = 180$.

  Note that X × X = C, V × X = L, and V × II = X.

Problems. Multiply the Roman numerals.

(a) LII × VIII =

(b) XXV × XV =

(c) LX × XII =

(d) CDXLIV × IX =

Long Division

Example. Perform long division.

$$
\begin{array}{r}
X \\
\hline
VII|CXXVI
\end{array}
\quad
\begin{array}{r}
X \\
\hline
VII|CXXVI \\
-LXX \\
\hline
LVI
\end{array}
\quad
\begin{array}{r}
XVIII \\
\hline
VII|CXXVI \\
-LXX \\
\hline
LVI \\
LVI
\end{array}
\quad
\begin{array}{r}
18 \\
\hline
7\,|\,126 \\
-70 \\
\hline
56 \\
56
\end{array}
$$

Notes: VII × X = LXX, CXXVI − LXX = LVI, and VII × VIII = LVI.

Problems. Perform long division.

(a)

$$IV\,|\,\overline{CCLII}$$

(b)

$$VI\,|\,\overline{CDL}$$

(c)

$$IX\,|\,\overline{DCCLVI}$$

(d)

$$XV\,|\,\overline{CDLXXX}$$

Remainders

The answer to a division problem includes a remainder when one number doesn't divide evenly into another number. For example, $14 \div 4 = 3R2$ (three with a remainder of two) since $4 \times 3 = 12$ and $14 - 12 = 2$.

Examples. Divide the Roman numerals.

- XXIII $\div$ V = XX $\div$ V + R III = IV R III.

  Check: $23 \div 5 = 20 \div 5 + R3 = 4R3$. Note: $5 \times 4 = 20$.

- LX $\div$ VII = LVI $\div$ VII + R IV = VIII R IV.

  Check: $60 \div 7 = 56 \div 7 + R4 = 8R4$. Note: $7 \times 8 = 56$.

Problems. Divide the Roman numerals.

(a) X $\div$ III =

(b) XXXV $\div$ IV =

(c) XX $\div$ VI =

(d) LV $\div$ IX =

(e) LI $\div$ VII =

(f) XL $\div$ IX =

(g) L $\div$ VIII =

Long Division with Remainders

Example. Perform long division.

$$
\begin{array}{r} XXX \\ VI\,\overline{|\,CCXV} \end{array}
\qquad
\begin{array}{r} XXX \\ VI\,\overline{|\,CCXV} \\ -CLXXX \\ \hline XXXV \end{array}
\qquad
\begin{array}{r} XXX \\ VI\,\overline{|\,CCXV} \\ -CLXXX \\ \hline XXXV \\ -XXX \\ \hline V \end{array}
\qquad
\begin{array}{r} XXXV\ R\ V \\ VI\,\overline{|\,CCXV} \\ -CLXXX \\ \hline XXXV \\ \end{array}
\qquad
\begin{array}{r} 35R5 \\ 6\,\overline{|\,215} \\ -180 \\ \hline 35 \\ -30 \\ \hline 5 \end{array}
$$

Notes: VI × XXX = CLXXX and CCXV − CLXXX = XXXV.

Problems. Perform long division.

(a)

$$IV\,\overline{|\,CCCLXVI}$$

(b)

$$VII\,\overline{|\,DXXI}$$

(c)

$$IX\,\overline{|\,DLVI}$$

(d)

$$XL\,\overline{|\,MC}$$

V

Powers and Roots

Chapter goals:

- Square and cube Roman numerals.
- Raise Roman numerals to various powers.
- Find the positive square roots of Roman numerals.
- Find cube and other roots of Roman numerals.

Square Roman Numerals

A number that has an exponent of two (Roman numeral II) is said to be squared. This means to multiply the number by itself. For example, V^{II} means $V \times V$.

Example. $V^{II} = V \times V = XXV$ ($5^2 = 5 \times 5 = 25$).

Problems. Square the Roman numerals.

(a) $III^{II} =$

(b) $VI^{II} =$

(c) $II^{II} =$

(d) $VIII^{II} =$

(e) $IV^{II} =$

(f) $X^{II} =$

(g) $XV^{II} =$

(h) $IX^{II} =$

(i) $VII^{II} =$

(j) $L^{II} =$

Cube Roman Numerals

A number that has an exponent of three (Roman numeral III) is said to be cubed. This means to multiply the number by itself three times. For example, V^{III} means $V \times V \times V$.

Example. $VI^{III} = VI \times VI \times VI = CCXVI$ ($6^3 = 6 \times 6 \times 6 = 216$).

Problems. Cube the Roman numerals.

(a) $II^{III} =$

(b) $V^{III} =$

(c) $III^{III} =$

(d) $X^{III} =$

(e) $IV^{III} =$

(f) $I^{III} =$

(g) $XI^{III} =$

(h) $VIII^{III} =$

(i) $IX^{III} =$

(j) $VII^{III} =$

Powers

In the expression b^p, the base b is raised to the power of p. The power p is called an exponent. For an exponent that is a whole number, this means to multiply the base b by itself p times. For example, V^{IV} means $V \times V \times V \times V$.

Example. $V^{IV} = V \times V \times V \times V = DCXXV$.

$\quad$ Check: $5^4 = 5 \times 5 \times 5 \times 5 = 25 \times 25 = 625$.

Problems. Evaluate each power.

$\quad$ (a) $II^{IV} =$

$\quad$ (b) $III^{IV} =$

$\quad$ (c) $II^{V} =$

$\quad$ (d) $IV^{IV} =$

$\quad$ (e) $II^{X} =$

$\quad$ (f) $III^{V} =$

$\quad$ (g) $V^{V} =$

$\quad$ (h) $VII^{IV} =$

$\quad$ (i) $II^{IX} =$

Square Roots

When a number is inside of a square root, like $\sqrt{XVI}$, this means, "What number squared is equal to XVI?" Since IV^{II} $= IV \times IV = XVI$, it follows that $\sqrt{XVI} = IV$.

Example. $\sqrt{XXXVI} = VI$ ($\sqrt{36} = 6$ since $6^2 = 6 \times 6 = 36$).

Problems. Find the positive square roots.

(a) $\sqrt{IV} =$

(b) $\sqrt{XXV} =$

(c) $\sqrt{I} =$

(d) $\sqrt{IX} =$

(e) $\sqrt{C} =$

(f) $\sqrt{LXIV} =$

(g) $\sqrt{DCXXV} =$

(h) $\sqrt{CXXI} =$

(i) $\sqrt{CCLXXXIX} =$

(j) $\sqrt{CD} =$

Roots

A small number placed to the left of a root symbol, like $\sqrt[n]{p}$, means, "What number multiplied by itself n times is equal to p?" For example, $\sqrt[III]{VIII}$ means, what number multiplied by itself three times is equal to VIII?

Example. $\sqrt[III]{VIII} = \text{II}$ ($\sqrt[3]{8} = 2$ since $2^3 = 2 \times 2 \times 2 = 8$).

Problems. Find the positive, real roots.

    (a) $\sqrt[III]{LXIV} =$

    (b) $\sqrt[IX]{DXII} =$

    (c) $\sqrt[III]{DXII} =$

    (d) $\sqrt[VI]{LXIV} =$

    (e) $\sqrt[IV]{DCXXV} =$

    (f) $\sqrt[VIII]{CCLVI} =$

    (g) $\sqrt[V]{CCXLIII} =$

    (h) $\sqrt[IV]{MCCXCVI} =$

    (i) $\sqrt[XI]{MMXLVIII} =$

VI
Fractions and Percents

Chapter goals:

- Make and interpret fractions with Roman numerals.
- Reduce fractions and make mixed numbers with Roman numerals.
- Add, subtract, multiply, and divide fractions with Roman numerals.
- Convert between fractions and percents.

Fractions with Roman Numerals

Problems. Express the fraction of each pie that is shaded using Roman numerals. The first is done as an example.

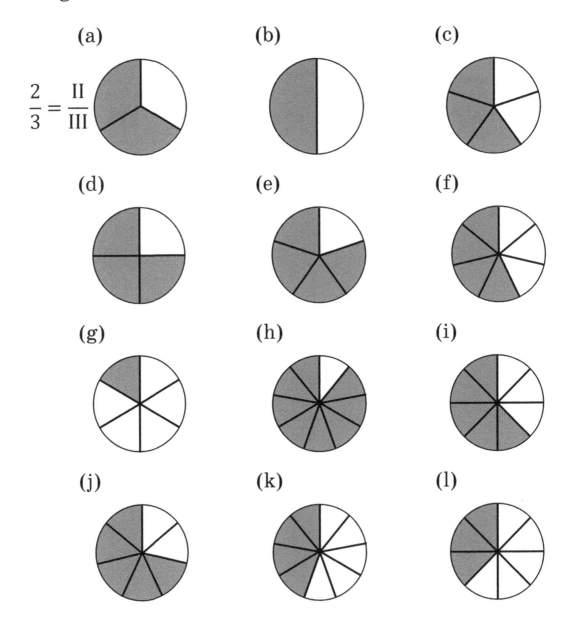

Reduced Fractions

If the numerator and denominator share a common factor, divide the numerator and denominator each by the greatest common factor. For example, $\dfrac{8}{12} = \dfrac{8 \div 4}{12 \div 4} = \dfrac{2}{3}$.

Example. $\dfrac{\text{XVIII}}{\text{XXIV}} = \dfrac{\text{XVIII} \div \text{VI}}{\text{XXIV} \div \text{VI}} = \dfrac{\text{III}}{\text{IV}} \left(\dfrac{18}{24} = \dfrac{18 \div 6}{24 \div 6} = \dfrac{3}{4} \right)$.

Problems. Reduce each fraction to its simplest form.

(a) $\dfrac{\text{X}}{\text{XV}} =$

(b) $\dfrac{\text{XII}}{\text{XXXVI}} =$

(c) $\dfrac{\text{C}}{\text{CD}} =$

(d) $\dfrac{\text{CD}}{\text{M}} =$

(e) $\dfrac{\text{V}}{\text{XC}} =$

(f) $\dfrac{\text{XIV}}{\text{XLIX}} =$

(g) $\dfrac{\text{L}}{\text{C}} =$

(h) $\dfrac{\text{CDLXXV}}{\text{MCM}} =$

Mixed Numbers

A mixed number includes an integer plus a fraction, like $3\frac{4}{5} = 3 + \frac{4}{5}$. A mixed number can be expressed as an improper fraction as follows: Multiply the integer and denominator, and add the numerator to make the new numerator, while the denominator remains unchanged: $3\frac{4}{5} = \frac{3\times5+4}{5} = \frac{19}{5}$. (The "improper" form is often preferred in math and science.)

Example. $V\frac{III}{X} = \frac{V\times X+III}{X} = \frac{LIII}{X} \left(5\frac{3}{10} = \frac{5\times10+3}{10} = \frac{53}{10}\right).$

Problems. Convert each number to an improper fraction.

(a) $II\frac{IV}{V} =$

(b) $III\frac{V}{VIII} =$

(c) $VI\frac{I}{IV} =$

(d) $I\frac{II}{III} =$

(e) $X\frac{VII}{L} =$

(f) $VIII\frac{VII}{IX} =$

Add and Subtract Fractions

To add or subtract fractions, multiply the numerator and denominator of each fraction by the factor needed to make the lowest common denominator.

Example. $\dfrac{V}{VIII} - \dfrac{VII}{XII} = \dfrac{V \times III}{VIII \times III} - \dfrac{VII \times II}{XII \times II} = \dfrac{XV}{XXIV} - \dfrac{XIV}{XXIV} = \dfrac{XV-XIV}{XXIV} = \dfrac{I}{XXIV}.$

Check: $\dfrac{5}{8} - \dfrac{7}{12} = \dfrac{5 \times 3}{8 \times 3} - \dfrac{7 \times 2}{12 \times 2} = \dfrac{15}{24} - \dfrac{14}{24} = \dfrac{15-14}{24} = \dfrac{1}{24}.$

Problems. Add or subtract the fractions.

(a) $\dfrac{I}{III} + \dfrac{I}{VI} =$

(b) $\dfrac{V}{IX} - \dfrac{I}{II} =$

(c) $\dfrac{I}{V} + \dfrac{III}{IV} =$

(d) $\dfrac{III}{IV} - \dfrac{II}{III} =$

(e) $\dfrac{VII}{X} + \dfrac{IV}{XV} =$

(f) $\dfrac{XI}{XII} - \dfrac{V}{VIII} =$

(g) $V - \dfrac{III}{IV} =$

Multiply Fractions

To multiply fractions, multiply the numerators together to make the new numerator and multiply the denominators together to make the new denominator.

Example. $\dfrac{II}{III} \times \dfrac{VI}{V} = \dfrac{II \times VI}{III \times V} = \dfrac{XII}{XV} = \dfrac{XII \div III}{XV \div III} = \dfrac{IV}{V}.$

Check: $\dfrac{2}{3} \times \dfrac{6}{5} = \dfrac{2 \times 6}{3 \times 5} = \dfrac{12}{15} = \dfrac{12 \div 3}{15 \div 3} = \dfrac{4}{5}.$

Problems. Multiply the fractions.

(a) $\dfrac{I}{II} \times \dfrac{V}{III} =$

(b) $\dfrac{III}{IV} \times \dfrac{VIII}{IX} =$

(c) $\dfrac{V}{VIII} \times \dfrac{IV}{XXV} =$

(d) $\dfrac{VII}{X} \times \dfrac{L}{XLIX} =$

(e) $\dfrac{VI}{LV} \times \dfrac{XXII}{III} =$

(f) $\dfrac{VII}{L} \times \dfrac{C}{XXI} =$

(g) $\dfrac{III}{M} \times \dfrac{CD}{IX} =$

Reciprocals

To find the reciprocal of an improper fraction, swap the numerator and denominator. For example, the reciprocal of $\frac{2}{3}$ equals $\frac{3}{2}$. To find the reciprocal of an integer, divide one by the integer. For example, the reciprocal of 4 equals $\frac{1}{4}$.

Example. The reciprocal of $\frac{II}{VII}$ equals $\frac{VII}{II}$ (or $III\frac{I}{II}$).

Problems. Find the reciprocal of each number.

(a) The reciprocal of $\frac{V}{VI}$ equals

(b) The reciprocal of $\frac{III}{VIII}$ equals

(c) The reciprocal of V equals

(d) The reciprocal of $\frac{IX}{IV}$ equals

(e) The reciprocal of $\frac{V}{II}$ equals

(f) The reciprocal of $\frac{I}{L}$ equals

(g) The reciprocal of $\frac{XI}{XC}$ equals

Divide Fractions

To divide one fraction by another, multiply the first fraction by the reciprocal of the second fraction. For example, the division $\frac{II}{V} \div \frac{VIII}{III}$ is equivalent to the multiplication $\frac{II}{V} \times \frac{III}{VIII}$.

Example. $\frac{II}{V} \div \frac{VIII}{III} = \frac{II}{V} \times \frac{III}{VIII} = \frac{II \times III}{V \times VIII} = \frac{VI}{XL} = \frac{VI \div II}{XL \div II} = \frac{III}{XX}.$

Check: $\frac{2}{5} \div \frac{8}{3} = \frac{2}{5} \times \frac{3}{8} = \frac{2 \times 3}{5 \times 8} = \frac{6}{40} = \frac{6 \div 2}{40 \div 2} = \frac{3}{20}.$

Problems. Divide the fractions.

(a) $\frac{I}{II} \div \frac{III}{IV} =$

(b) $\frac{II}{III} \div \frac{IV}{IX} =$

(c) $\frac{III}{X} \div \frac{VII}{V} =$

(d) $\frac{IV}{VII} \div \frac{V}{II} =$

(e) $\frac{V}{VI} \div \frac{I}{III} =$

(f) $\frac{III}{L} \div \frac{I}{C} =$

(g) $\frac{IX}{D} \div \frac{XI}{M} =$

Percents

A percent is a fraction of one hundred. (The term "percent" refers to a specific value, like 25%, whereas "percentage" refers to a general amount, like "a percentage of the toys.") A percent can be expressed as a fraction by dividing by 100 and (if applicable) reducing the fraction.

Example. $LX\% = \frac{LX}{C} = \frac{LX \div XX}{C \div XX} = \frac{III}{V} \left(60\% = \frac{60}{100} = \frac{60 \div 20}{100 \div 20} = \frac{3}{5} \right).$

Problems. Convert each percent to a reduced fraction.

(a) $L\% =$

(b) $XXV\% =$

(c) $LXXX\% =$

(d) $CL\% =$

(e) $CD\% =$

(f) $XV\% =$

(g) $LXIV\% =$

Notes on Decimals and Large Numbers

If you naively try to write down a decimal value with Roman numerals, you will run into a problem with ambiguity. For example, X.II could be read as 10.2 or as 10.11: How can you tell whether the two I's form the single digit II = 2 or if they are separate digits I = 1 and I = 1? Similarly, does X.IV mean 10.4 or does it mean 10.15? So if your heart desires to make decimal values with Roman numerals, you need a way to separate the digits (like X.I-II-IV for 10.124). Of course, the Romans didn't use decimals. For fractional units, the Romans preferred base twelve (not base ten like decimals), since 2, 3, and 4 factor into 12 (but not 10). Another problem with decimals is that there is no Roman numeral for zero.

Have you wondered about large numbers? How do you make a number larger than 3999? There isn't a standard. One way is to use a bar to multiply by a thousand: $\overline{V} = 5000$. If you are familiar with scientific notation, you could multiply by a power of ten: $CDXVII \times X^{IV} = 417 \times 10^4 = 4{,}170{,}000$.

VII

Number Pattern Puzzles

Chapter goals:

- Identify a variety of patterns in sequences which consist of Roman numerals.

- Predict the Roman numerals that come next in a sequence.

- Apply logic and reasoning skills.

Set I (Easy)

Puzzles. Predict the next two Roman numerals.

(a) IX, XVII, XXV, XXXIII,

(b) LXIV, LII, XL, XXVIII,

(c) XVII, XXXIV, LII, LXXI,

(d) CXLIV, CXIX, XCIV, LXIX,

(e) VI, XI, XVIII, XXVII,

(f) CCLXXIV, CCXIV, CLXVI, CXXX,

(g) MMMLXXII, DCCLXVIII, CXCII, XLVIII,

(h) I, V, X, L, C,

(i) MMCLVI, MDCCXXXIII, MCCCX, DCCCLXXXVII,

(j) I, II, VI, XXIV,

Set II (Easy)

Puzzles. Predict the next two Roman numerals.

(a) XXV, XXXVI, XLIX, LXIV,

(b) IV, V, IX, XIV, XXIII,

(c) X, XIX, XXXI, XLVI,

(d) C, CC, CD, DCCC,

(e) CLXXII, CVII, LXV, XLII, XXIII,

(f) IV, VI, X, XVIII, XXXIV, LXVI,

(g) DCCXX, CCCLX, CXX, XXX,

(h) DXII, CCCXLIII, CCXVI, CXXV,

(i) MDCCXCII, DCCCXCVI, CDXLVIII, CCXXIV,

(j) XVII, XIX, XXIII, XXIX, XXXI, XXXVII,

Set III (Medium)

Puzzles. Predict the next two Roman numerals.

(a) IV, XVI, XL, LXXXVIII, CLXXXIV,

(b) V, MM, X, M, XX, D, XL, CCL, LXXX,

(c) V, X, XX, XL, L, C, CX, CCXX,

(d) II, V, XI, XVII, XXIII, XXXI, XLI, XLVII,

(e) IV, XVI, VIII, XXXII, XVI, LXIV, XXXII,

(f) X, CX, XX, C, XL, XC, LXXX, LXXX, CLX, LXX

(g) VX, VL, IL, LC, VC, IC,

(h) XIV, XVI, XLI, XLV, LIV, LVI, LIX,

(i) II, III, VI, IX, XXXVI, XLI, CCXLVI,

(j) VII, L, XV, XLIII, XXIII, XXXVI, XXXI, XXIX, XXXIX,

Set IV (Hard)

Puzzles. Predict the next two Roman numerals.

(a) III, VII, XII, XIV, XVI, XIX, XXI,

(b) X, XXIX, LXVI, CXXVII, CCXVIII,

(c) I, II, IV, VII, XIII, XXIV, XLIV, LXXXI,

(d) IV, VI, X, XIV, XXII, XXVI, XXXIV, XXXVIII, XLVI,

(e) II, III, XIX, XX, XXX, CXC, CC,

(f) II, III, VI, XI, XX, XXXV, LX, CI, CLXVIII,

(g) III, IV, VI, VIII, XII, XIV, XVIII, XX, XXIV,

(h) VIII, XLIX, CCXVI, DCXXV, MXXIV,

(i) IV, IX, IX, XIII, XIV, XVII, XIX, XXI, XXIV, XXV,

(j) CXXI, CLXIX, CCLXXXIX, CCCLXI, DXXIX, DCCCXLI,

Set V (II×II Arrays)

Puzzles. Fill in the missing element of each array.

(a)

IX	III
XV	VIII

XII	IV
XVI	VII

L	X
C	XV

C	L
D	

(b)

II	III
V	IV

IV	VI
IX	XII

III	V
X	XV

V	X
L	

(c)

III	IV
II	XX

IV	VI
L	V

XL	II
V	VI

X	
V	I

(d)

V	II
III	I

VI	III
IV	VI

IX	IV
VI	XV

C	V
L	

Set VI (III×III Arrays)

Puzzles. Fill in the missing elements of each array.

(a)

II	IV	VI
XXVI	XVI	X
		CX

(b)

LXIV		
XVI	C	CCLVI
IV	XXXVI	CXLIV

(c)

		XCIV
CDXC	CDIV	XLIX
	XCIX	XLIV

(d)

		I
	XI	II
XXXVII	VII	IV

Set VII (I+II Pyramids)

Puzzles. Fill in the missing element of each pyramid.

(a)

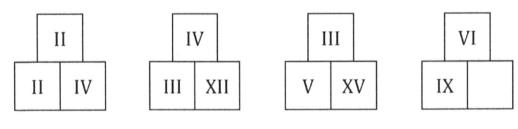

(b)

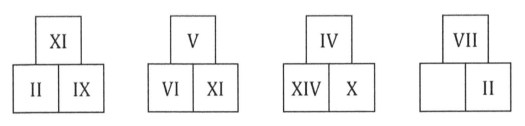

(c)

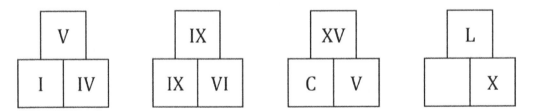

(d)

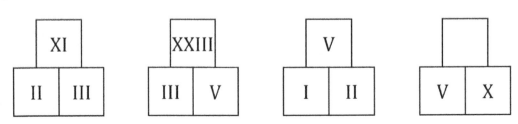

Set VIII (I+II+III Pyramids)

Puzzles. Fill in the missing elements of each pyramid.

(a)

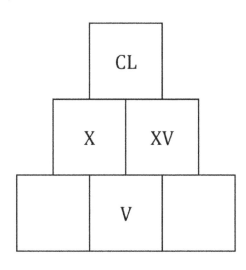

(b)

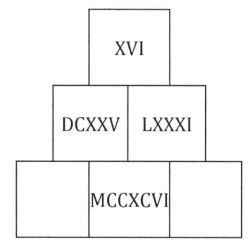

(c)

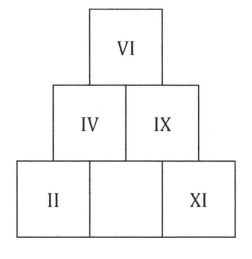

(d)

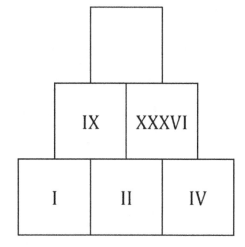

Set IX (I+II+III+IV+V Pyramid)

Puzzle. Fill in the missing elements of the pyramid.

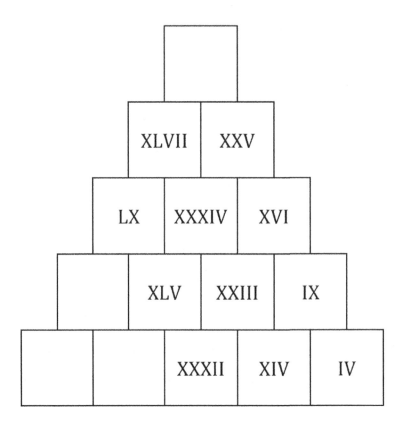

VIII
Variety Puzzles

Chapter goals:

- Apply Roman numerals to a variety of mathematical puzzles.
- Solve a few word puzzles that have a Roman theme.
- Have fun.

Roman Numeral Builder

Puzzle. How many regular Roman numerals can you make using only the letters in the word below? List them in the space below. Rule: In a single Roman numeral, don't use a letter more times than it appears in the word below.

C–I–V–I–L

Word Builder

Puzzle. How many words can you make using only the basic Roman numerals below? List them in the space below. This time, you may use each Roman numeral as many times as you want in a single word.

I–V–X–L–C–D–M

Roman Numeral Fill-in

Puzzle. Fill in the grid below with Roman numerals, using the list of ordinary numbers at the bottom of the page. Use each Roman numeral exactly one time, one letter per box.

13	87	249
28	88	286
38	169	350
85	175	685

750	1850
1085	1975
1676	2600
1705	2980

Cryptograms

Puzzles. Decode each cryptogram below.

(a)   YKV AGUGQQVDL KEB DFBVPZPGDFB

OEQQEZVQ EFB YPEO BGGPQ YKEY

AGDUB TV DQVB RGP VUETGPEYV

QOVAWEU VRRVAYQ.

(b)   DIHDBOO FWL W DIDXTWA ABLIAC SIA

FBWTCKE AIHWQL XQCOT OC FWL

JBLCAIEBJ GE CKB MITNWQI, HIXQC

MBLXMOXL.

Roman Numeral Sudoku

Puzzles. Enter the Roman numerals I thru VI into the grid such that no Roman numeral is repeated in the same row, column, or bold rectangle.

(a)

		V	I		VI
III			IV		II
	I			V	
II		I			
	III				I
			III	II	

(b)

				III	V
		IV		I	
II			III		IV
		III		V	VI
I	II	IV			
V				IV	

(c)

	II	VI		V	
VI			V		IV
III		II			
		I		III	
V					VI
	VI		III		

(d)

	V			II	
			VI		I
VI		III	II		
I		II			V
		II		V	I
		III	IV		

Roman Numeral Kakuro

Puzzle. Enter the Roman numerals I thru IX into the grid such that no Roman numeral is repeated in any horizontal or vertical block, and such that the Roman numerals in each horizontal block add up to the sum shown to its left and such that the Roman numerals in each vertical block add up to the sum shown above it. Tip: Begin by looking for sums that must have all very large or all very small digits. The solution involves logic similar to Sudoku solutions.

Word Scrambles

Puzzles. Unscramble each Roman-themed word. When you finish, rearrange the bubbled letters to form a bonus word.

(a)   E D U Q U A C T

(b)   S C A B A U

(c)   M E R O P E R

(d)   U S S T L Y

(e)   C H O O T R

(f)   T H O C I R R E

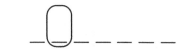

(g)   R U N E C T Y

(h)   Z I N C T I E

(i)   The bonus word is: _ _ _ _ _ _ _ _

Word Search

Puzzle. Find the listed words in the puzzle.

```
Y T R Q U M S E B I R C S K V Q J W
L E S I H C O T G I O R O X U L Z D
K U B H I E R O G L Y P H I C S A I
N Q U P Z E H W O V A R I R P U L M
J N O R T A P S N L G D D H Q P E A
G A X C M F S M A S M O I E P M X R
L B E B A E R C O M U N Y A O Y A Y
A W H Q U T E S O P X T T M T L N P
T N S M S M N N K R S O U A X O D S
S M E V O C U G T B I L M R H F R U
E A B V L M E Q I R H Z L T B H I Q
D P E X E G O D A E C I N A P I A G
E D H N U L W H O A R A H P T Y L O
P L T Y M R C S P I K K J O L I G N
G K I H T N I X A R F Q V E X T N E
J Z O N R A S E A C S U I L U J W K
M O R T E L P M E T Y N O C L A F V
```

ALEXANDRIA	HIEROGLYPHICS	PATRON
BANQUET	JULIUS CAESAR	PEDESTAL
BRUTUS	LATIN	PHARAOH
CHARIOT	LUXOR	POMPEII
CHISEL	MARK ANTHONY	PYRAMID
CLEOPATRA	MAUSOLEUM	REIGN
COLOSSEUM	MONUMENT	SCRIBE
EGYPT	NILE	SPHINX
FALCON	OLYMPUS	TEMPLE
GLADIATOR	PALACE	THEBES

Roman Numeral Math Crossword

Puzzle. Answer each clue using Roman numerals, writing a single letter in a box.

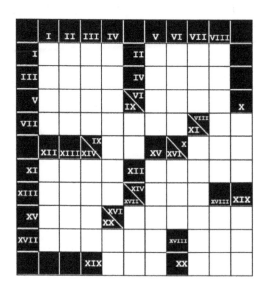

ACROSS

I. MMM − C   II. CXC × VI

III. MM − CCC   IV. MC − XXX

V. ML ÷ III   VI. XXX − I

VII. DC − XI   VIII. $\sqrt{CD}$

IX. I + II + III   X. DIII × II

XI. MIV ÷ IV   XII. MD − XCVI

XIII. DX + CI   XIV. XII^{II} − IV

XV. V × VI   XVI. C − XI

XVII. L − II   XVIII. XX − I

XIX. II^{IV} XX. $\sqrt{XLIX}$

DOWN

I. L^{II} − C   II. CL × XI

III. CD − XC   IV. MCC − XXXIII

V. MX + MI   VI. CC − XXX

VII. III × VII   VIII. C − XIV

IX. XX − IX   X. II × VII

XI. XV × CV   XII. D − LXXX

XIII. VI × XL   XIV. CL ÷ II

XV. MIX + CIII   XVI. III × XL

XVII. VI × IX   XVIII. $\sqrt{IX}$

XIX. CV ÷ V   XX. $\sqrt{LXXXI}$

IX
Word Problems

Chapter goals:

- Solve word problems that involve Roman numerals.

- Apply logic, reasoning, and arithmetic skills.

- Since the ancient Romans didn't have calculators, you should solve these problems by hand. (However, you are encouraged to use an abacus.)

- Solve word problems with a Roman theme.

Lifetime Problems

(a) Roman emperor Augustus was born on September XXIII in LXIII BC and died on August XIX in XIV AD. Note that the year "zero" did not exist (the calendar skips from I BC to I AD). What was Augustus' age in years when he died?

(b) Atticus and Cecilia were twins born on May IX in CXLVII BC. Atticus lived for LIX years and III months, while Cecilia lived for LXI years and VIII months. Determine the year of each twin's death.

Comparison Problems

(a) Felix and Cyrus originally each had the same number of grapes. After Felix received XXIV grapes from Cyrus, Felix had III times as many grapes as Cyrus. How many grapes did each boy have after Cyrus gave those grapes to Felix?

(b) Flavia and Octavia each have rooms in the shape of a square. Octavia's room is IV pedes (Roman feet) wider than Flavia's room, and the area of Octavia's room is LXXX pedes quadratum (Roman square feet) larger than Flavia's room. How wide is each girl's bedroom?

Age Comparison Problems

(a) Atticus is currently III times as old as Diana. IV years ago, Atticus was V times as old as Diana. What are their ages now?

(b) Priscilla is currently X years older than Magnus. V years ago, Priscilla was VI times as old as Magnus. What are their ages now?

Purchase Problems

(a) A Roman soldier buys a pair of boots for XCIV denarii and a pair of sandals for LXIX denarii. The soldier pays with two argenti. One argentus is equal in value to C denarii. How many denarii should the soldier receive back in change (if there is no sales tax)? Note: argenti is the plural of argentus.

(b) A farmer buys IV cows and VI goats for ML denarii at a place where the price of a cow is twice the price of a goat. How much did the farmer pay for each cow and each goat (if there is no sales tax)?

Percent Problems

(a) A customer buys clothing. The prices add up to MMCM denarii. There is sales tax of I percent. What total amount must the customer pay?

(b) A customer buys a table. The price of the table is MCCL denarii. The customer negotiates a discount of XX percent, but the customer must pay sales tax of half a percent. How much must the customer pay for the table, in total?

Fraction Problems

(a) A school has LXXXI students. XXXVI of the students are boys. What is the ratio of girls to boys at the school?

(b) A mom bakes a pie. Her daughter eats $\frac{I}{VI}$ of the pie. Her son eats $\frac{II}{V}$ of what remains? What fraction of the pie is left?

Conversion Problems

A few units of measure used by ancient Romans included the pes (one Roman foot, for which the plural is pedes), the stadium (equal to DCXXV pedes), the hora (equal to a twelfth of the daylight hours, with a plural of horae), the uncia (a Roman ounce), and the libra (a pound, equal to XII unciae).

(a) The perimeter of the Colosseum is approximately equal to III stadia. What is the perimeter in pedes?

(b) If a chariot weighed DCCXX unciae, what was its weight in librae? (Note: librae is the plural of libra.)

Rate Problems

(a) Claudia and Valentina are each running XII laps. Each lap measures XX stadia. Claudia runs XL stadia per hora, while Valentina runs XXX stadia per hora. Once Claudia finishes, how long will she need to wait for Valentina to finish?

(b) Marcus and Julius are initially separated by ML stadia. At the same time, Marcus begins riding a chariot with an average speed of CL stadia per hora directly towards Julius and Julius begins riding a horse with an average speed of CC stadia per hora directly towards Marcus. About when will Marcus and Julius meet?

Team Problems

(a) Camilla, Octavia, and Aurelia need to bake LXXX cakes. They each have their own kitchens. Camilla can bake V cakes per hora, Octavia can bake VII cakes per hora, and Aurelia can bake VIII cakes per hora. How long will it take the three women to bake all of the cakes?

(b) Lucius and Titus are building a brick wall. Lucius could finish the wall in III horae if he worked alone. Titus could finish the wall in VI horae if he worked alone. How long will it take them to finish the wall if they work together?

ANSWER KEY

I – Converting to Roman Numerals

Page VII – Less Than Forty

      (a) $30 = 10 + 10 + 10 = $ XXX

      (b) $15 = 10 + 5 = $ XV

      (c) $3 = 1 + 1 + 1 = $ III

      (d) $21 = 10 + 10 + 1 = $ XXI

      (e) $19 = 10 + 9 = 10 + (10 - 1) = $ XIX

      (f) $37 = 10 + 10 + 10 + 5 + 1 + 1 = $ XXXVII

      (g) $28 = 10 + 10 + 5 + 1 + 1 + 1 = $ XXVIII

      (h) $12 = 10 + 1 + 1 = $ XII

      (i) $36 = 10 + 10 + 10 + 5 + 1 = $ XXXVI

      (j) $24 = 10 + 10 + 4 = 10 + 10 + (5 - 1) = $ XXIV

Page VIII – Forty to One Hundred

      (a) $72 = 50 + 20 + 2 = $ LXXII

      (b) $58 = 50 + 8 = $ LVIII

      (c) $67 = 50 + 10 + 7 = $ LXVII

      (d) $45 = 40 + 5 = (50 - 10) + 5 = $ XLV (although VL may seem tempting, this would be irregular, as mentioned at the end of the chapter)

      (e) $91 = 90 + 1 = (100 - 10) + 1 = $ XCI

      (f) $83 = 50 + 30 + 3 = $ LXXXIII

      (g) $49 = 40 + 9 = (50 - 10) + (10 - 1) = $ XLIX (note that IL is irregular, as mentioned at the end of the chapter)

      (h) $64 = 50 + 10 + 4 = 50 + 10 + (5 - 1) = $ LXIV

      (i) $86 = 50 + 30 + 6 = $ LXXXVI

      (j) $99 = 90 + 9 = (100 - 10) + (10 - 1) = $ XCIX (note that IC is irregular, as mentioned at the end of the chapter)

Page IX – One Hundred to Four Hundred

       (a) $110 = 100 + 10 = $ CX

       (b) $328 = 300 + 20 + 8 = $ CCCXXVIII

       (c) $222 = 200 + 20 + 2 = $ CCXXII

       (d) $286 = 200 + 50 + 30 + 6 = $ CCLXXXVI

       (e) $136 = 100 + 30 + 6 = $ CXXXVI

       (f) $300 = 100 + 100 + 100 = $ CCC

       (g) $375 = 300 + 50 + 20 + 5 = $ CCCLXXV

       (h) $194 = 100 + 90 + 4 = 100 + (100 - 10) + (5 - 1) = $ CXCIV

       (i) $150 = 100 + 50 = $ CL

       (j) $349 = 300 + 40 + 9 = 300 + (50 - 10) + (10 - 1) = $ CCCXLIX (note that CCCIL is irregular, as mentioned at the end of the chapter)

Page X – Four Hundred to One Thousand

       (a) $638 = 500 + 100 + 30 + 8 = $ DCXXXVIII

       (b) $810 = 500 + 300 + 10 = $ DCCCX

       (c) $984 = 900 + 84 = (1000 - 100) + 50 + 30 + 4 = $ CMLXXXIV

       (d) $492 = 400 + 90 + 2 = (500 - 100) + (100 - 10) + 2 = $ CDXCII

       (e) $549 = 500 + 40 + 9 = 500 + (50 - 10) + 9 = $ DXLIX (note that DIL is irregular, as mentioned at the end of the chapter)

       (f) $517 = 500 + 10 + 7 = $ DXVII

       (g) $839 = 500 + 300 + 30 + 9 = $ DCCCXXXIX

       (h) $464 = 400 + 64 = (500 - 100) + 50 + 10 + 4 = $ CDLXIV

       (i) $777 = 500 + 200 + 50 + 20 + 7 = $ DCCLXXVII

       (j) $999 = 900 + 90 + 9 = (1000 - 100) + (100 - 10) + 9 = $ CMXCIX (note that IM is irregular, as mentioned at the end of the chapter)

Page XI – One Thousand to Four Thousand

       (a) $1714 = 1000 + 500 + 200 + 10 + 4 = $ MDCCXIV

       (b) $2625 = 2000 + 500 + 100 + 20 + 5 = $ MMDCXXV

       (c) $3050 = 3000 + 50 = $ MMML

       (d) $1111 = 1000 + 100 + 10 + 1 = $ MCXI

       (e) $2027 = 2000 + 20 + 7 = $ MMXXVII

(f) $3289 = 3000 + 200 + 50 + 30 + 9 = $ MMMCCLXXXIX

(g) $3921 = 3000 + 900 + 20 + 1 = $ MMMCMXXI

(h) $2343 = 2000 + 300 + 40 + 3 = $ MMCCCXLIII

(i) $1568 = 1000 + 500 + 50 + 10 + 8 = $ MDLXVIII

(j) $3498 = 3000 + 400 + 90 + 8 = $ MMMCDXCVIII (note that MMMIID is irregular, as mentioned at the end of the chapter)

Page XII – Additional Practice

(a) $35 = 30 + 5 = $ XXXV

(b) $29 = 20 + 9 = $ XXIX

(c) $44 = 40 + 4 = $ XLIV

(d) $61 = 50 + 10 + 1 = $ LXI

(e) $88 = 50 + 30 + 8 = $ LXXXVIII

(f) $90 = 100 - 10 = $ XC

(g) $97 = 90 + 7 = $ XCVII (note that IIIC is irregular, as mentioned at the end of the chapter)

(h) $113 = 100 + 10 + 3 = $ CXIII

(i) $126 = 100 + 20 + 6 = $ CXXVI

(j) $172 = 100 + 50 + 20 + 2 = $ CLXXII

(k) $201 = 200 + 1 = $ CCI

(l) $218 = 200 + 10 + 8 = $ CCXVIII

(m) $350 = 300 + 50 = $ CCCL

(n) $387 = 300 + 50 + 30 + 7 = $ CCCLXXXVII

(o) $484 = 400 + 50 + 30 + 4 = $ CDLXXXIV

(p) $555 = 500 + 50 + 5 = $ DLV

(q) $719 = 500 + 200 + 10 + 9 = $ DCCXIX

(r) $926 = 900 + 20 + 6 = $ CMXXVI

(s) $1023 = 1000 + 20 + 3 = $ MXXIII

(t) $1242 = 1000 + 200 + 40 + 2 = $ MCCXLII

(u) $1897 = 1000 + 500 + 300 + 90 + 7 = $ MDCCCXCVII

(v) $1974 = 1000 + 900 + 50 + 20 + 4 = $ MCMLXXIV

(w) $2456 = 2000 + 400 + 50 + 6 = $ MMCDLVI

(x) 2735 = 2000 + 500 + 200 + 30 + 5 = MMDCCXXXV

(y) 2999 = 2000 + 900 + 90 + 9 = MMCMXCIX (note that MMIM is irregular, as mentioned at the end of the chapter)

(z) 3418 = 3000 + 400 + 10 + 8 = MMMCDXVIII

Page XIII – Irregular Additions

(a) 9 = "VIIII" (irregular) and IX (regular)

(b) 400 = "CCCC" (irregular) and CD (regular)

(c) 90 = "LXXXX" (irregular) and XC (regular)

(d) 1042 = "MXXXXII" (irregular) and MXLII (regular)

(e) 29 = "XXVIIII" (irregular) and XXIX (regular)

(f) 494 = "CCCCLXXXXIIII" (irregular) and CDXCIV (regular)

(g) 3900 = "MMMDCCCC" (irregular) and MMMCM (regular)

Page XIV – Irregular Subtractions

(a) 495 = "VD" should be 400 + 90 + 5 = CDXCV

(b) 549 = "DIL" should be 500 + 40 + 9 = DXLIX

(c) 1950 = "MLM" should be 1000 + 900 + 50 = MCML

(d) 145 = "CVL" should be 100 + 40 + 5 = CXLV

(e) 450 = "LD" should be 400 + 50 = CDL

(f) 990 = "XM" should be 900 + 90 = CMXC

(g) 499 = "ID" should be 400 + 90 + 9 = CDXCIX

Note: A few of these might be considered "mistakes" more than "irregular."

II – Converting from Roman Numerals

Page XVII – Less Than Forty

(a) XXIII = 10 + 10 + 3 = 23

(b) VI = 5 + 1 = 6

(c) XVII = 10 + 5 + 2 = 17

(d) XIV = 10 + (5 − 1) = 10 + 4 = 14

(e) XXV = 10 + 10 + 5 = 25

(f) XXXI = 10 + 10 + 10 + 1 = 31

(g) XXXVIII = 10 + 10 + 10 + 5 + 3 = 38

(h) XX $= 10 + 10 = 20$

(i) XXIV $= 10 + 10 + (5 - 1) = 20 + 4 = 24$

(j) XXXIX $= 10 + 10 + 10 + (10 - 1) = 30 + 9 = 39$

Page XVIII – Forty to One Hundred

(a) LXVIII $= 50 + 10 + 5 + 3 = 68$

(b) LV $= 50 + 5 = 55$

(c) XLVII $= (50 - 10) + 5 + 2 = 40 + 7 = 47$

(d) LXXIII $= 50 + 10 + 10 + 3 = 73$

(e) XC $= 100 - 10 = 90$

(f) XLIV $= (50 - 10) + (5 - 1) = 40 + 4 = 44$

(g) LIII $= 50 + 3 = 53$

(h) LXXXVII $= 50 + 10 + 10 + 10 + 5 + 2 = 87$

(i) LXXVI $= 50 + 10 + 10 + 5 + 1 = 76$

(j) XCIV $= (100 - 10) + (5 - 1) = 90 + 4 = 94$

Page XIX – One Hundred to Four Hundred

(a) CLXVI $= 100 + 50 + 10 + 5 + 1 = 166$

(b) CCXXXVIII $= 200 + 30 + 5 + 3 = 238$

(c) CXCIII $= 100 + (100 - 10) + 3 = 100 + 90 + 3 = 193$

(d) CCCLXV $= 300 + 50 + 10 + 5 = 365$

(e) CCLIV $= 200 + 50 + (5 - 1) = 250 + 4 = 254$

(f) CXXIX $= 100 + 20 + (10 - 1) = 120 + 9 = 129$

(g) CCXLIX $= 200 + (50 - 10) + (10 - 1) = 200 + 40 + 9 = 249$

(h) CCLXXXII $= 200 + 50 + 30 + 2 = 282$

(i) CXVI $= 100 + 10 + 5 + 1 = 116$

(j) CCCXCIV $= 300 + (100 - 10) + (5 - 1) = 300 + 90 + 4 = 394$

Page XX – Four Hundred to One Thousand

(a) DCLXI $= 500 + 100 + 50 + 10 + 1 = 661$

(b) DCCCXXXIII $= 500 + 300 + 30 + 3 = 833$

(c) CDLXII $= (500 - 100) + 50 + 10 + 2 = 400 + 62 = 462$

(d) CMXCI $= (1000 - 100) + (100 - 10) + 1 = 900 + 90 + 1 = 991$

(e) DCCXIV $= 500 + 200 + 10 + (5 - 1) = 710 + 4 = 714$

(f) DLXXVII = 500 + 50 + 20 + 5 + 2 = 577

(g) CMXXIX = (1000 − 100) + 20 + (10 − 1) = 900 + 20 + 9 = 929

(h) DCCLX = 500 + 200 + 50 + 10 = 760

(i) DV = 500 + 5 = 505

(j) CDXLIV = (500 − 100) + (50 − 10) + (5 − 1) = 400 + 40 + 4 = 444

Page XXI – One Thousand to Four Thousand

(a) MDCLXVI = 1000 + 500 + 100 + 50 + 10 + 6 = 1666

(b) MMMCCIII = 3000 + 200 + 3 = 3203

(c) MMMCDLV = 3000 + 400 + 50 + 5 = 3455

(d) MMCMXXII = 2000 + 900 + 20 + 2 = 2922

(e) MMCLXXII = 2000 + 100 + 50 + 20 + 2 = 2172

(f) MMMDCCCXLI = 3000 + 500 + 300 + 40 + 1 = 3841

(g) MCMXIX = 1000 + 900 + 10 + 9 = 1919

(h) MCCCLXXXIII = 1000 + 300 + 50 + 30 + 3 = 1383

(i) MMDXLVII = 2000 + 500 + 40 + 7 = 2547

(j) MMMCMXCIX = 3000 + 900 + 90 + 9 = 3999

Page XXII – Additional Practice

(a) VIII = 5 + 3 = 8

(b) IX = 10 − 1 = 9

(c) XXVI = 20 + 5 + 1 = 26

(d) XLII = (50 − 10) + 2 = 42

(e) LXVI = 50 + 10 + 5 + 1 = 66

(f) LXXXV = 50 + 30 + 5 = 85

(g) XC = 100 − 10 = 90

(h) CIV = 100 + (5 − 1) = 104

(i) CXL = 100 + (50 − 10) = 140

(j) CCXXXII = 200 + 30 + 2 = 232

(k) CCCXCI = 300 + (100 − 10) + 1 = 300 + 90 + 1 = 391

(l) CDXLIX = (500 − 100) + (50 − 10) + (10 − 1) = 400 + 40 + 9 = 449

(m) DLIII = 500 + 50 + 3 = 553

(n) DCVI = 500 + 100 + 5 + 1 = 606

(o) DCCL = 500 + 200 + 50 = 750

(p) DCCCXCVII = 500 + 300 + (100 − 10) + 5 + 2 = 800 + 90 + 7 = 897

(q) CMXIV = (1000 − 100) + 10 + (5 − 1) = 900 + 10 + 4 = 914

(r) MXI = 1000 + 10 + 1 = 1011

(s) MCCXLII = 1000 + 200 + (50 − 10) + 2 = 1200 + 40 + 2 = 1242

(t) MCDXV = 1000 + (500 − 100) + 10 + 5 = 1000 + 400 + 15 = 1415

(u) MMXX = 2000 + 20 = 2020

(v) MMDCCL = 2000 + 500 + 200 + 50 = 2750

(w) MMCMVI = 2000 + (1000 − 100) + 5 + 1 = 2000 + 900 + 6 = 2906

(x) MMMCLXVI = 3000 + 100 + 50 + 10 + 5 + 1 = 3166

(y) MMMCDXL = 3000 + (500 − 100) + (50 − 10) = 3000 + 400 + 40 = 3440

(z) MMMCMLIX = 3000 + (1000 − 100) + 50 + 9 = 3000 + 900 + 59 = 3959

Page XXIII – Irregular Additions

(a) "VVV" = 5 + 5 + 5 = 15 should be 10 + 5 = XV

(b) "LXXXX" = 50 + 40 = 90 should be 100 − 10 = XC

(c) "DD" = 500 + 500 = 1000 should be 1000 = M

(d) "CCCCC" = 100 × 5 = 500 should be 500 = D

(e) "CLLIIII" = 100 + 50 + 50 + 4 = 204 should be 200 + 4 = CCIV

(f) "DCCCCV" = 500 + 400 + 5 = 905 should be 900 + 5 = CMV

(g) "XXVVII" = 10 + 10 + 5 + 5 + 1 + 1 = 32 should be 30 + 2 = XXXII

Note: Some of these would be considered "mistakes" more than "irregular."

Page XXIV – Irregular Subtractions

(a) "VL" = 50 − 5 = 45 should be 40 + 5 = XLV

(b) "IC" = 100 − 1 = 99 should be 90 + 9 = XCIX

(c) "XD" = 500 − 10 = 490 should be 400 + 90 = CDXC

(d) "CIL" = 100 + (50 − 1) = 149 should be 100 + 40 + 9 = CXLIX

(e) "MID" = 1000 + (500 − 1) = 1499 should be 1000 + 400 + 90 + 9 = MCDXCIX

(f) "DLC" = 500 + (100 − 50) = 550 should be 500 + 50 = DL

(g) "MLD" = 1000 + (500 − 50) = 1450 should be 1000 + 400 + 50 = MCDL

Note: Some of these would be considered "mistakes" more than "irregular."

III – Reading and Writing Roman Numerals

Page XXVI – Roman Numerals in Use

(a) IV $= 5 - 1 = 4$

(b) XIV $= 10 + (5 - 1) = 10 + 4 = 14$ (not the only King Louis)

(c) XXVI $= 10 + 10 + 5 + 1 = 26$ (each game has its own number)

(d) VII $= 5 + 2 = 7$ (this is just one of many King Henry's; if Henry the eighth came to your mind faster, for example, maybe he is now more widely known than Henry the seventh)

(e) xxxiv $= 10 + 10 + 10 + (5 - 1) = 30 + 4 = 34$ (introductory page numbers are often written with lowercase Roman numerals)

(f) XI $= 10 + 1 = 11$ (eleven o'clock)

(g) XLIX $= (50 - 10) + (10 - 1) = 40 + 9 = 49$ (note that IL would have been irregular)

Page XXVII – Years

(a) $753 = 500 + 200 + 50 + 3 =$ DCCLIII BC

(b) $80 = 50 + 30 =$ LXXX AD

(c) $509 = 500 + (10 - 1) =$ DIX BC

(d) $1372 = 1000 + 300 + 50 + 20 + 2 =$ MCCCLXXII AD

(e) $410 = (500 - 100) + 10 =$ CDX AD

(f) $1483 = 1000 + (500 - 100) + 50 + 30 + 3 =$ MCDLXXXIII AD

(g) $1564 = 1000 + 500 + 50 + 10 + 4 =$ MDLXIV AD

Page XXVIII – Lifetimes

(a) LXIII $= 50 + 10 + 3 = 63$ BC to XIV $= 10 + (5 - 1) = 10 + 4 = 14$ AD

(b) CXI $= 100 + 10 + 1 = 111$ BC to LXXI $= 50 + 20 + 1 = 71$ BC

(c) CCLXXII $= 200 + 50 + 22 = 272$ AD to CCCXXXVII $= 300 + 37 = 337$ AD

(d) CVI $= 100 + 5 + 1 = 106$ BC to XLVIII $= (50 - 10) + 5 + 3 = 48$ BC

(e) CCCLVI $= 300 + 50 + 6$ BC $= 356$ BC to CCCXXIII $= 300 + 23 = 323$ BC

(f) MCLXII $= 1000 + 100 + 50 + 12 = 1162$ AD to
MCCXXVII $= 1000 + 200 + 20 + 7 = 1227$ AD

Page XXIX – U.S. Holidays

(a) July 4, 1777 = July IV, MDCCLXXVII (since $4 = 5 - 1 = $ IV

and since $1777 = 1000 + 500 + 200 + 50 + 20 + 7 = $ MDCCLXXVII)

Note: Are you wondering why it's 1777 instead of 1776? It's because the first "holiday" was celebrated a year after the historic event occurred in 1776

(b) May 30, 1868 = May XXX, MDCCCLXVIII (since $30 = 10 + 10 + 10 = $ XXX

and since $1868 = 1000 + 500 + 300 + 50 + 10 + 8 = $ MDCCCLXVIII)

(c) November 11, 1919 = November XI, MCMXIX (since $11 = 10 + 1 = $ XI

and since $1919 = 1000 + 900 + 10 + 9 = $ MCMXIX)

(d) December 31, 1999 = December XXXI, MCMXCIX (since $31 = 30 + 1 = $ XXXI

and since $1999 = 1000 + 900 + 90 + 9 = $ MCMXCIX)

(e) December 25, 1849 = December XXV, MDCCCXLIX (since $25 = 20 + 5 = $ XXV

and since $1849 = 1000 + 500 + 300 + 40 + 9 = $ MDCCCXLIX)

(f) October 31, 2094 = October XXXI, MMXCIV (since $31 = 30 + 1 = $ XXXI

and since $2094 = 2000 + 90 + 4 = $ MMXCIV)

Page XXX – Historical Dates

(a) August 28, 1963 (since XXVIII $= 20 + 8 = 28$

and since MCMLXIII $= 1000 + 900 + 50 + 10 + 3 = 1963$)

(b) November 9, 1799 (since IX $= 10 - 1 = 9$

and since MDCCXCIX $= 1000 + 500 + 200 + 90 + 9 = 1799$)

(c) December 2, 1804 (since II $= 1 + 1 = 2$

and since MDCCCIV $= 1000 + 500 + 300 + 4 = 1804$)

(d) June 15, 1215 (since XV $= 10 + 5 = 15$

and since MCCXV $= 1000 + 200 + 10 + 5 = 1215$)

(e) December 25, 800 (since XXV $= 20 + 5 = 25$

and since DCCC $= 500 + 300 = 800$)

(f) December 16, 1773 (since XVI $= 10 + 6 = 16$

and since MDCCLXXIII $= 1000 + 500 + 200 + 50 + 20 + 3 = 1773$)

IV – Arithmetic with Roman Numerals

Page XXXII – Addition and Subtraction Facts

(a) II + II = IV (2 + 2 = 4)   (b) VII − V = II (7 − 5 = 2)

(c) VIII − III = V (8 − 3 = 5)   (d) IV + I = V (4 + 1 = 5)

(e) VI − IV = II (6 − 4 = 2)   (f) VI + V = XI (6 + 5 = 11)

(g) VII + II = IX (7 + 2 = 9)   (h) V − I = IV (5 − 1 = 4)

(i) X − II = VIII (10 − 2 = 8)   (j) VIII + VI = XIV (8 + 6 = 14)

(k) V + V = X (5 + 5 = 10)   (l) XVI − VII = IX (16 − 7 = 9)

(m) III + III = VI (3 + 3 = 6)   (n) X − III = VII (10 − 3 = 7)

(o) XI − IX = II (11 − 9 = 2)   (p) V + IV = IX (5 + 4 = 9)

(q) IX + VIII = XVII (9 + 8 = 17)   (r) XV − VII = VIII (15 − 7 = 8)

(s) X − I = IX (10 − 1 = 9)   (t) IX + VI = XV (9 + 6 = 15)

(u) XI − IV = VII (11 − 4 = 7)   (v) VI + III = IX (6 + 3 = 9)

(w) VIII + VII = XV (8 + 7 = 15)   (x) IX − IV = V (9 − 4 = 5)

(y) IX + V = XIV (9 + 5 = 14)   (z) XIV − V = IX (14 − 5 = 9)

Page XXXIII – Adding Larger Roman Numerals

(a) XVI + XII = XX + V + III = $\boxed{\text{XXVIII}}$ (16 + 12 = 28)

(b) XXXVIII + XXI = XXX + XX + V + III + I = L + V + IV = L + IX = $\boxed{\text{LIX}}$
(38 + 21 = 59) Notes: XXX + XX = L, III + I = IV, and V + IV = IX

(c) XXXV + XV = XXX + X + V + V = XXX + X + X = $\boxed{\text{L}}$ (35 + 15 = 50)

(d) LXIV + LXVI = L + L + XX + (V − I) + V + I = C + XX + V + V
= C + XX + X = $\boxed{\text{CXXX}}$ (64 + 66 = 130) Notes: L + L = C and −I + I = 0

(e) CLXVI + LXV = C + L + L + XX + V + V + I = C + C + XX + X + I
= $\boxed{\text{CCXXXI}}$ (166 + 65 = 231) Notes: L + L = C and V + V = X

(f) XCIV + XLIII = (C − X) + (V − I) + (L − X) + III = C + L − XX + V − I + III
= C + L − XX + V + II = C + XXX + V + II = $\boxed{\text{CXXXVII}}$ (94 + 43 = 137)
Notes: L − XX = XXX and −I + III = II

(g) CMLXXXIII + CDLXXII = (M − C) + (D − C) + L + L + XXX + XX + III + II
= M + D − CC + L + L + XXX + XX + III + II = M + D − CC + C + L + V
= M + D − C + L + V = $\boxed{\text{MCDLV}}$ (983 + 472 = 1455)
Notes: L + L = C, XXX + XX = L, III + II = V, −CC + C = −C and D − C = CD

Page XXXIV – Subtracting Larger Roman Numerals

(a) XXXVIII − XXVI = XXX + V + III − (XX + V + I)

= XXX + V + III − XX − V − I = X + II = $\boxed{\text{XII}}$ (38 − 26 = 12) Note: V − V = 0

(b) XXXIV − XV = XXX + (V − I) − (X + V) = XXX + V − I − X − V

= XX − I = $\boxed{\text{XIX}}$ (34 − 15 = 19) Note: V − V = 0

(c) LXXVIII − LXI = L + XX + V + III − (L + X + I)

= L + XX + V + III − L − X − I = X + V + II = $\boxed{\text{XVII}}$ (78 − 61 = 17)

Notes: L − L = 0, XX − X = X, and III − I = II

(d) LXXII − XXVI = L + XX + II − (XX + V + I) = L + XX + II − XX − V − I

= L − V + I = XLV + I = $\boxed{\text{XLVI}}$ (72 − 26 = 46) Note: L − V = XLV (50 − 5 = 45)

(e) C − LXV = C − (L + X + V) = C − L − X − V

= L + XXX + X + V + V − L − X − V = XXX + V = $\boxed{\text{XXXV}}$ (100 − 65 = 35)

Note: C = L + XXX + X + V + V (100 = 50 + 30 + 10 + 5 + 5)

(f) CMXL − CDV = (M − C) + (L − X) − (D − C + V)

= M − C + L − X − D + C − V = M − D + L − X − V = D + XL − V

= D + XXXV = $\boxed{\text{DXXXV}}$ (940 − 405 = 535) Notes: −(−C) = +C

Distribute the minus sign: −(D − C + V) = −D − (−C) − V = −D + C − V

L − X = XL, LX − V = XXXV (40 − 5 = 35), and M − D = D

Page XXXV – Multiplication Facts

(a) II × III = VI (2 × 3 = 6)   (b) II × II = IV (2 × 2 = 4)

(c) IV × IV = XVI (4 × 4 = 16)   (d) III × III = IX (3 × 3 = 9)

(e) VI × IV = XXIV (6 × 4 = 24)   (f) V × II = X (5 × 2 = 10)

(g) VII × III = XXI (7 × 3 = 21)   (h) VIII × IV = XXXII (8 × 4 = 32)

(i) I × II = II (1 × 2 = 2)   (j) VIII × VI = XLVIII (8 × 6 = 48)

(k) V × V = XXV (5 × 5 = 25)   (l) IX × II = XVIII (9 × 2 = 18)

(m) VIII × VIII = LXIV (8 × 8 = 64) (n) VII × II = XIV (7 × 2 = 14)

(o) VI × V = XXX (6 × 5 = 30)   (p) IX × IV = XXXVI (9 × 4 = 36)

(q) IX × VIII = LXXII (9 × 8 = 72) (r) VII × VII = XLIX (7 × 7 = 49)

(s) VI × VI = XXXVI (6 × 6 = 36)   (t) IX × VII = LXIII (9 × 7 = 63)

(u) VII × VI = XLII (7 × 6 = 42)   (v) VIII × III = XXIV (8 × 3 = 24)

(w) IX × VI = LIV (9 × 6 = 54)   (x) VII × IV = XXVIII (7 × 4 = 28)

(y) VIII × VII = LVI (8 × 7 = 56)   (z) IX × IX = LXXXI (9 × 9 = 81)

Page XXXVI – Division Facts

(a) $V \div V = I$ $(5 \div 5 = 1)$ (b) $XV \div V = III$ $(15 \div 5 = 3)$

(c) $XII \div III = IV$ $(12 \div 3 = 4)$ (d) $XL \div VIII = V$ $(40 \div 8 = 5)$

(e) $XLIX \div VII = VII$ $(49 \div 7 = 7)$ (f) $XXXII \div VIII = IV$ $(32 \div 8 = 4)$

(g) $XXXVI \div IX = IV$ $(36 \div 9 = 4)$ (h) $XXVII \div IX = III$ $(27 \div 9 = 3)$

(i) $VIII \div II = IV$ $(8 \div 2 = 4)$ (j) $LXIII \div VII = IX$ $(63 \div 7 = 9)$

(k) $XLVIII \div VI = VIII$ $(48 \div 6 = 8)$ (l) $LXIV \div VIII = VIII$ $(64 \div 8 = 8)$

(m) $LVI \div VII = VIII$ $(56 \div 7 = 8)$ (n) $LIV \div VI = IX$ $(54 \div 6 = 9)$

(o) $XVI \div IV = IV$ $(16 \div 4 = 4)$ (p) $XXXVI \div VI = VI$ $(36 \div 6 = 6)$

(q) $LXXII \div VIII = IX$ $(72 \div 8 = 9)$ (r) $LXIII \div IX = VII$ $(63 \div 9 = 7)$

(s) $XLII \div VI = VII$ $(42 \div 6 = 7)$ (t) $XXV \div V = V$ $(25 \div 5 = 5)$

(u) $LIV \div IX = VI$ $(54 \div 9 = 6)$ (v) $LVI \div VIII = VII$ $(56 \div 8 = 7)$

(w) $XX \div IV = V$ $(20 \div 4 = 5)$ (x) $XLII \div VII = VI$ $(42 \div 7 = 6)$

(y) $XLVIII \div VIII = VI$ $(48 \div 8 = 6)$ (z) $LXXII \div IX = VIII$ $(72 \div 9 = 8)$

Page XXXVII – Multiplying Larger Roman Numerals

(a) $LII \times VIII = (L + II) \times VIII = L \times VIII + II \times VIII = CD + XVI = \boxed{CDXVI}$

Check: $52 \times 8 = (50 + 2) \times 8 = 50 \times 8 + 2 \times 8 = 400 + 16 = 416$

Note: $L \times VIII = CD$ $(50 \times 8 = 400)$

(b) $XXV \times XV = (XX + V) \times (X + V) = XX \times X + XX \times V + V \times X + V \times V$

$= CC + C + L + XXV = \boxed{CCCLXXV}$

Check: $25 \times 15 = (20 + 5) \times 15 = 20 \times 15 + 5 \times 15 = 300 + 75 = 375$

Notes: $XX \times X = CC$ $(20 \times 10 = 200)$ and $XX \times V = C$ $(20 \times 5 = 100)$

(c) $LX \times XII = (L + X) \times XII = L \times XII + X \times XII = DC + CXX = \boxed{DCCXX}$

Check: $60 \times 12 = (60) \times (10 + 2) = 60 \times 10 + 60 \times 2 = 600 + 120 = 720$

Notes: $L \times XII = DC$ $(50 \times 12 = 600)$ and $X \times XII = CXX$ $(10 \times 12 = 120)$

(d) $CDXLIV \times IX = (CD + XL + IV) \times IX = CD \times IX + XL \times IX + IV \times IX$

$= MMMDC + CCCLX + XXXVI = \boxed{MMMCMXCVI}$

Check: $444 \times 9 = (400 + 40 + 4) \times 9 = 400 \times 9 + 40 \times 9 + 4 \times 9$

$= 3600 + 360 + 36 = 3996$ Notes: $CD \times IX = MMMDC$ $(400 \times 9 = 3600)$

$DC + CCC = CM$ $(600 + 300 = 900)$, and $LX + XXX = XC$ $(60 + 30 = 90)$

Page XXXVIII – Long Division

(a) Check: $4 \times 63 = 252$

$$
\begin{array}{r} \\ \text{IV}\,\overline{)\,\text{CCLII}} \end{array}
\qquad
\begin{array}{r} \text{LX} \\ \text{IV}\,\overline{)\,\text{CCLII}} \\ -\text{CCXL} \end{array}
\qquad
\begin{array}{r} \text{LX} \\ \text{IV}\,\overline{)\,\text{CCLII}} \\ -\text{CCXL} \\ \hline \text{XII} \end{array}
\qquad
\begin{array}{r} \text{LXIII} \\ \text{IV}\,\overline{)\,\text{CCLII}} \\ -\text{CCXL} \\ \hline \text{XII} \\ \text{XII} \end{array}
\qquad
\begin{array}{r} 63 \\ 4\,\overline{)\,252} \\ -240 \\ \hline 12 \\ 12 \end{array}
$$

(b) Check: $6 \times 75 = 450$

$$
\begin{array}{r} \\ \text{VI}\,\overline{)\,\text{CDL}} \end{array}
\qquad
\begin{array}{r} \text{LXX} \\ \text{VI}\,\overline{)\,\text{CDL}} \\ -\text{CDXX} \end{array}
\qquad
\begin{array}{r} \text{LXX} \\ \text{VI}\,\overline{)\,\text{CDL}} \\ -\text{CDXX} \\ \hline \text{XXX} \end{array}
\qquad
\begin{array}{r} \text{LXXV} \\ \text{VI}\,\overline{)\,\text{CDL}} \\ -\text{CDXX} \\ \hline \text{XXX} \\ \text{XXX} \end{array}
\qquad
\begin{array}{r} 75 \\ 6\,\overline{)\,450} \\ -420 \\ \hline 30 \\ 30 \end{array}
$$

(c) Check: $9 \times 84 = 756$

$$
\begin{array}{r} \\ \text{IX}\,\overline{)\,\text{DCCLVI}} \end{array}
\qquad
\begin{array}{r} \text{LXXX} \\ \text{IX}\,\overline{)\,\text{DCCLVI}} \\ -\text{DCCXX} \end{array}
\qquad
\begin{array}{r} \text{LXXX} \\ \text{IX}\,\overline{)\,\text{DCCLVI}} \\ -\text{DCCXX} \\ \hline \text{XXXVI} \end{array}
\qquad
\begin{array}{r} \text{LXXXIV} \\ \text{IX}\,\overline{)\,\text{DCCLVI}} \\ -\text{DCCXX} \\ \hline \text{XXXVI} \\ \text{XXXVI} \end{array}
\qquad
\begin{array}{r} 84 \\ 9\,\overline{)\,756} \\ -720 \\ \hline 36 \\ 36 \end{array}
$$

(d) Check: $15 \times 32 = 480$

$$
\begin{array}{r} \\ \text{XV}\,\overline{)\,\text{CDLXXX}} \end{array}
\qquad
\begin{array}{r} \text{XXX} \\ \text{XV}\,\overline{)\,\text{CDLXXX}} \\ -\text{CDL} \end{array}
\qquad
\begin{array}{r} \text{XXX} \\ \text{XV}\,\overline{)\,\text{CDLXXX}} \\ -\text{CDL} \\ \hline \text{XXX} \end{array}
\qquad
\begin{array}{r} \text{XXXII} \\ \text{XV}\,\overline{)\,\text{CDLXXX}} \\ -\text{CDL} \\ \hline \text{XXX} \\ \text{XXX} \end{array}
\qquad
\begin{array}{r} 32 \\ 15\,\overline{)\,480} \\ -450 \\ \hline 30 \\ 30 \end{array}
$$

Page XXXIX – Remainders

(a) X ÷ III = IX ÷ III + R I = III R I (10 ÷ 3 = 3R1)

Check: 3 × 3 = 9 and R = 10 − 9 = 1

(b) XXXV ÷ IV = XXXII ÷ IV + R III = VIII R III (35 ÷ 4 = 8R3)

Check: 4 × 8 = 32 and R = 35 − 32 = 3

(c) XX ÷ VI = XVIII ÷ VI + R II = III R II (20 ÷ 6 = 3R2)

Check: 6 × 3 = 18 and R = 20 − 18 = 2

(d) LV ÷ IX = LIV ÷ IX + R I = VI R I (55 ÷ 9 = 6R1)

Check: 9 × 6 = 54 and R = 55 − 54 = 1

(e) LI ÷ VII = XLIX ÷ VII + R II = VII R II (51 ÷ 7 = 7R2)

Check: 7 × 7 = 49 and R = 51 − 49 = 2

(f) XL ÷ IX = XXXVI ÷ IX + R IV = IV R IV (40 ÷ 9 = 4R4)

Check: 9 × 4 = 36 and R = 40 − 36 = 4

(g) L ÷ VIII = XLVIII ÷ VIII + R II = VI R II (50 ÷ 8 = 6R2)

Check: 8 × 6 = 48 and R = 50 − 48 = 2

Page XL – Long Division with Remainders

(a) Check: 4 × 91 = 364 and R = 366 − 364 = 2

```
                  XC          XC          XCI R II       91R2
IV│CCCLXVI  IV│CCCLXVI IV│CCCLXVI IV│CCCLXVI  4│366
              −CCCLX      −CCCLX      −CCCLX      −360
                            VI          VI          6
                                       −IV         −4
                                        II          2
```

(b) Check: 7 × 74 = 518 and R = 521 − 518 = 3

```
                  LXX         LXX         LXXIVRIII     74R3
VII│DXXI   VII│DXXI    VII│DXXI    VII│DXXI    7│521
              −CDXC       −CDXC       −CDXC       −490
                            XXXI        XXXI        31
                                       −XXVIII     −28
                                         III        3
```

(c) Check: $9 \times 61 = 549$ and $R = 556 - 549 = 7$

$$\begin{array}{c} \\ \text{IX} \overline{\big)\text{DLVI}} \end{array} \qquad \begin{array}{c} \text{LX}\\ \text{IX} \overline{\big)\text{DLVI}}\\ -\text{DXL} \end{array} \qquad \begin{array}{c} \text{LX}\\ \text{IX} \overline{\big)\text{DLVI}}\\ -\text{DXL}\\ \hline \text{XVI} \end{array} \qquad \begin{array}{c} \text{LXI R VII}\\ \text{IX} \overline{\big)\text{DLVI}}\\ -\text{DXL}\\ \hline \text{XVI}\\ -\text{IX}\\ \hline \text{VII} \end{array} \qquad \begin{array}{c} \text{61R7}\\ 9 \overline{\big)556}\\ -540\\ \hline 16\\ -9\\ \hline 7 \end{array}$$

(d) Check: $40 \times 27 = 1080$ and $R = 1100 - 1080 = 20$

$$\begin{array}{c} \\ \text{XL} \overline{\big)\text{MC}} \end{array} \qquad \begin{array}{c} \text{XX}\\ \text{XL} \overline{\big)\text{MC}}\\ -\text{DCCC} \end{array} \qquad \begin{array}{c} \text{XX}\\ \text{XL} \overline{\big)\text{MC}}\\ -\text{DCCC}\\ \hline \text{CCC} \end{array} \qquad \begin{array}{c} \text{XXVII R XX}\\ \text{XL} \overline{\big)\text{MC}}\\ -\text{DCCC}\\ \hline \text{CCC}\\ -\text{CCLXXX}\\ \hline \text{XX} \end{array} \qquad \begin{array}{c} \text{27R20}\\ 40 \overline{\big)1100}\\ -800\\ \hline 300\\ -280\\ \hline 20 \end{array}$$

V – Powers and Roots

Page XLII – Square Roman Numerals

(a) $\text{III}^{\text{II}} = \text{III} \times \text{III} = \text{IX}$ ($3^2 = 3 \times 3 = 9$)

(b) $\text{VI}^{\text{II}} = \text{VI} \times \text{VI} = \text{XXXVI}$ ($6^2 = 6 \times 6 = 36$)

(c) $\text{II}^{\text{II}} = \text{II} \times \text{II} = \text{IV}$ ($2^2 = 2 \times 2 = 4$)

(d) $\text{VIII}^{\text{II}} = \text{VIII} \times \text{VIII} = \text{LXIV}$ ($8^2 = 8 \times 8 = 64$)

(e) $\text{IV}^{\text{II}} = \text{IV} \times \text{IV} = \text{XVI}$ ($4^2 = 4 \times 4 = 16$)

(f) $\text{X}^{\text{II}} = \text{X} \times \text{X} = \text{C}$ ($10^2 = 10 \times 10 = 100$)

(g) $\text{XV}^{\text{II}} = \text{XV} \times \text{XV} = \text{CCXXV}$ ($15^2 = 15 \times 15 = 225$)

(h) $\text{IX}^{\text{II}} = \text{IX} \times \text{IX} = \text{LXXXI}$ ($9^2 = 9 \times 9 = 81$)

(i) $\text{VII}^{\text{II}} = \text{VII} \times \text{VII} = \text{XLIX}$ ($7^2 = 7 \times 7 = 49$)

(j) $\text{L}^{\text{II}} = \text{L} \times \text{L} = \text{MMD}$ ($50^2 = 50 \times 50 = 2500$)

Page XLIII – Cube Roman Numerals

(a) $II^{III} = II \times II \times II = VIII$ ($2^3 = 2 \times 2 \times 2 = 4 \times 2 = 8$)

(b) $V^{III} = V \times V \times V = CXXV$ ($5^3 = 5 \times 5 \times 5 = 25 \times 5 = 125$)

(c) $III^{III} = III \times III \times III = XXVII$ ($3^3 = 3 \times 3 \times 3 = 9 \times 3 = 27$)

(d) $X^{III} = X \times X \times X = M$ ($10^3 = 10 \times 10 \times 10 = 100 \times 10 = 1000$)

(e) $IV^{III} = IV \times IV \times IV = LXIV$ ($4^3 = 4 \times 4 \times 4 = 16 \times 4 = 64$)

(f) $I^{III} = I \times I \times I = I$ ($1^3 = 1 \times 1 \times 1 = 1$)

(g) $XI^{III} = XI \times XI \times XI = MCCCXXXI$ ($11^3 = 11 \times 11 \times 11 = 121 \times 11 = 1331$)

(h) $VIII^{III} = VIII \times VIII \times VIII = DXII$ ($8^3 = 8 \times 8 \times 8 = 64 \times 8 = 512$)

(i) $IX^{III} = IX \times IX \times IX = DCCXXIX$ ($9^3 = 9 \times 9 \times 9 = 81 \times 9 = 729$)

(j) $VII^{III} = VII \times VII \times VII = CCCXLIII$ ($7^3 = 7 \times 7 \times 7 = 49 \times 7 = 343$)

Page XLIV – Powers

(a) $II^{IV} = II \times II \times II \times II = XVI$ ($2^4 = 2 \times 2 \times 2 \times 2 = 4 \times 4 = 16$)

(b) $III^{IV} = III \times III \times III \times III = LXXXI$ ($3^4 = 3 \times 3 \times 3 \times 3 = 9 \times 9 = 81$)

(c) $II^{V} = II \times II \times II \times II \times II = XXXII$ ($2^5 = 2 \times 2 \times 2 \times 2 \times 2 = 8 \times 4 = 32$)

(d) $IV^{IV} = IV \times IV \times IV \times IV = CCLVI$ ($4^4 = 4 \times 4 \times 4 \times 4 = 16 \times 16 = 256$)

(e) $II^{X} = II \times II \times II \times II \times II \times II \times II \times II \times II \times II = MXXIV$

($2^{10} = 2 \times 2 \times 2 \times 2 \times 2 \times 2 \times 2 \times 2 \times 2 \times 2 = 4 \times 4 \times 4 \times 4 \times 4 = 64 \times 16 = 1024$)

(f) $III^{V} = III \times III \times III \times III \times III = CCXLIII$ ($3^5 = 3 \times 3 \times 3 \times 3 \times 3 = 27 \times 9 = 243$)

(g) $V^{V} = V \times V \times V \times V \times V = MMMCXXV$ ($5^5 = 5 \times 5 \times 5 \times 5 \times 5 = 125 \times 25 = 3125$)

(h) $VII^{IV} = VII \times VII \times VII \times VII = MMCDI$ ($7^4 = 7 \times 7 \times 7 \times 7 = 49 \times 49 = 2401$)

(i) $II^{IX} = II \times II \times II \times II \times II \times II \times II \times II \times II = DXII$

($2^9 = 2 \times 2 \times 2 \times 2 \times 2 \times 2 \times 2 \times 2 \times 2 = 8 \times 8 \times 8 = 64 \times 8 = 512$)

Page XLV – Square Roots

(a) $\sqrt{IV} = II$ ($\sqrt{4} = 2$ since $2^2 = 2 \times 2 = 4$)

(b) $\sqrt{XXV} = V$ ($\sqrt{25} = 5$ since $5^2 = 5 \times 5 = 25$)

(c) $\sqrt{I} = I$ ($\sqrt{1} = 1$ since $1^2 = 1 \times 1 = 1$)

(d) $\sqrt{IX} = III$ ($\sqrt{9} = 3$ since $3^2 = 3 \times 3 = 9$)

(e) $\sqrt{C} = X$ ($\sqrt{100} = 10$ since $10^2 = 10 \times 10 = 100$)

(f) $\sqrt{LXIV} = VIII$ ($\sqrt{64} = 8$ since $8^2 = 8 \times 8 = 64$)

(g) $\sqrt{\text{DCXXV}} = \text{XXV}$ ($\sqrt{625} = 25$ since $25^2 = 25 \times 25 = 625$)

(h) $\sqrt{\text{CXXI}} = \text{XI}$ ($\sqrt{121} = 11$ since $11^2 = 11 \times 11 = 121$)

(i) $\sqrt{\text{CCLXXXIX}} = \text{XVII}$ ($\sqrt{289} = 17$ since $17^2 = 17 \times 17 = 289$)

(j) $\sqrt{\text{CD}} = \text{XX}$ ($\sqrt{400} = 20$ since $20^2 = 20 \times 20 = 400$)

Page XLVI – Roots

(a) $\sqrt[\text{III}]{\text{LXIV}} = \text{IV}$ ($\sqrt[3]{64} = 4$ since $4^3 = 4 \times 4 \times 4 = 16 \times 4 = 64$)

(b) $\sqrt[\text{IX}]{\text{DXII}} = \text{II}$ ($\sqrt[9]{512} = 2$ since $2^9 = 2 \times 2 \times 2 \times 2 \times 2 \times 2 \times 2 \times 2 \times 2$
$= 8 \times 8 \times 8 = 64 \times 8 = 512$)

(c) $\sqrt[\text{III}]{\text{DXII}} = \text{VIII}$ ($\sqrt[3]{512} = 8$ since $8^3 = 8 \times 8 \times 8 = 64 \times 8 = 512$)

(d) $\sqrt[\text{VI}]{\text{LXIV}} = \text{II}$ ($\sqrt[6]{64} = 2$ since $2^6 = 2 \times 2 \times 2 \times 2 \times 2 \times 2 = 8 \times 8 = 64$)

(e) $\sqrt[\text{IV}]{\text{DCXXV}} = \text{V}$ ($\sqrt[4]{625} = 5$ since $5^4 = 5 \times 5 \times 5 \times 5 = 25 \times 25 = 625$)

(f) $\sqrt[\text{VIII}]{\text{CCLVI}} = \text{II}$ ($\sqrt[8]{256} = 2$ since $2^8 = 2 \times 2 \times 2 \times 2 \times 2 \times 2 \times 2 \times 2$
$= 4 \times 4 \times 4 \times 4 = 16 \times 16 = 256$)

(g) $\sqrt[\text{V}]{\text{CCXLIII}} = \text{III}$ ($\sqrt[5]{243} = 3$ since $3^5 = 3 \times 3 \times 3 \times 3 \times 3 = 27 \times 9 = 243$)

(h) $\sqrt[\text{IV}]{\text{MCCXCVI}} = \text{VI}$ ($\sqrt[4]{1296} = 6$ since $6^4 = 6 \times 6 \times 6 \times 6 = 36 \times 36 = 1296$)

(i) $\sqrt[\text{XI}]{\text{MMXLVIII}} = \text{II}$ ($\sqrt[11]{2048} = 2$ since
$2^{11} = 2 \times 2 \times 2 \times 2 \times 2 \times 2 \times 2 \times 2 \times 2 \times 2 \times 2 = 8 \times 8 \times 8 \times 4 = 64 \times 32 = 2048$)

VI – Fractions and Percents

Page XLVIII – Fractions with Roman Numerals

(a) $\frac{2}{3} = \frac{\text{II}}{\text{III}}$

(b) $\frac{1}{2} = \frac{\text{I}}{\text{II}}$

(c) $\frac{3}{5} = \frac{\text{III}}{\text{V}}$

(d) $\frac{3}{4} = \frac{\text{III}}{\text{IV}}$

(e) $\frac{4}{5} = \frac{\text{IV}}{\text{V}}$

(f) $\frac{4}{7} = \frac{\text{IV}}{\text{VII}}$

(g) $\frac{1}{6} = \frac{\text{I}}{\text{VI}}$

(h) $\frac{8}{9} = \frac{\text{VIII}}{\text{IX}}$

(i) $\frac{5}{8} = \frac{\text{V}}{\text{VIII}}$

(j) $\frac{5}{7} = \frac{\text{V}}{\text{VII}}$

(k) $\frac{4}{9} = \frac{\text{IV}}{\text{IX}}$

(l) $\frac{3}{8} = \frac{\text{III}}{\text{VIII}}$

Page XLIX – Reduced Fractions

(a) $\dfrac{X}{XV} = \dfrac{X \div V}{XV \div V} = \dfrac{II}{III} \left(\dfrac{10}{15} = \dfrac{10 \div 5}{15 \div 5} = \dfrac{2}{3} \right)$

(b) $\dfrac{XII}{XXXVI} = \dfrac{XII \div XII}{XXXVI \div XII} = \dfrac{I}{III} \left(\dfrac{12}{36} = \dfrac{12 \div 12}{36 \div 12} = \dfrac{1}{3} \right)$ If you get $\dfrac{2}{6}, \dfrac{4}{9}$, or $\dfrac{6}{18}$, reduce further.

(c) $\dfrac{C}{CD} = \dfrac{C \div C}{CD \div C} = \dfrac{I}{IV} \left(\dfrac{100}{400} = \dfrac{100 \div 100}{400 \div 100} = \dfrac{1}{4} \right)$

(d) $\dfrac{CD}{M} = \dfrac{CD \div CC}{M \div CC} = \dfrac{II}{V} \left(\dfrac{400}{1000} = \dfrac{400 \div 200}{1000 \div 200} = \dfrac{2}{5} \right)$ If you get $\dfrac{4}{10}$, reduce further.

(e) $\dfrac{V}{XC} = \dfrac{V \div V}{XC \div V} = \dfrac{I}{XVIII} \left(\dfrac{5}{90} = \dfrac{5 \div 5}{90 \div 5} = \dfrac{1}{18} \right)$ Note that $90 \div 5 = 18$.

(f) $\dfrac{XIV}{XLIX} = \dfrac{XIV \div VII}{XLIX \div VII} = \dfrac{II}{VII} \left(\dfrac{14}{49} = \dfrac{14 \div 7}{49 \div 7} = \dfrac{2}{7} \right)$

(g) $\dfrac{L}{C} = \dfrac{L \div L}{C \div L} = \dfrac{I}{II} \left(\dfrac{50}{100} = \dfrac{50 \div 50}{100 \div 50} = \dfrac{1}{2} \right)$

(h) $\dfrac{CDLXXV}{MCM} = \dfrac{CDLXXV \div CDLXXV}{MCM \div CDLXXV} = \dfrac{I}{IV} \left(\dfrac{475}{1900} = \dfrac{475 \div 475}{1900 \div 475} = \dfrac{1}{4} \right)$

Page L – Mixed Numbers

(a) $II\dfrac{IV}{V} = \dfrac{II \times V + IV}{V} = \dfrac{XIV}{V} \left(2\dfrac{4}{5} = \dfrac{2 \times 5 + 4}{5} = \dfrac{14}{5} \right)$

(b) $III\dfrac{V}{VIII} = \dfrac{III \times VIII + V}{VIII} = \dfrac{XXIX}{VIII} \left(3\dfrac{5}{8} = \dfrac{3 \times 8 + 5}{8} = \dfrac{29}{8} \right)$

(c) $VI\dfrac{I}{IV} = \dfrac{VI \times IV + I}{IV} = \dfrac{XXV}{IV} \left(6\dfrac{1}{4} = \dfrac{6 \times 4 + 1}{4} = \dfrac{25}{4} \right)$

(d) $I\dfrac{II}{III} = \dfrac{I \times III + II}{III} = \dfrac{V}{III} \left(1\dfrac{2}{3} = \dfrac{1 \times 3 + 2}{3} = \dfrac{5}{3} \right)$

(e) $X\dfrac{VII}{L} = \dfrac{X \times L + VII}{L} = \dfrac{DVII}{L} \left(10\dfrac{7}{50} = \dfrac{10 \times 50 + 7}{50} = \dfrac{500 + 7}{50} = \dfrac{507}{50} \right)$

(f) $VIII\dfrac{VII}{IX} = \dfrac{VIII \times IX + VII}{IX} = \dfrac{LXXIX}{IX} \left(8\dfrac{7}{9} = \dfrac{8 \times 9 + 7}{9} = \dfrac{72 + 7}{9} = \dfrac{79}{9} \right)$

Page LI – Add and Subtract Fractions

(a) $\dfrac{I}{III} + \dfrac{I}{VI} = \dfrac{I \times II}{III \times II} + \dfrac{I \times I}{VI \times I} = \dfrac{II}{VI} + \dfrac{I}{VI} = \dfrac{II + I}{VI} = \dfrac{III}{VI} = \dfrac{III \div III}{VI \div III} = \dfrac{I}{II}$

Check: $\dfrac{1}{3} + \dfrac{1}{6} = \dfrac{1 \times 2}{3 \times 2} + \dfrac{1 \times 1}{6 \times 1} = \dfrac{2}{6} + \dfrac{1}{6} = \dfrac{2 + 1}{6} = \dfrac{3}{6} = \dfrac{3 \div 3}{6 \div 3} = \dfrac{1}{2}$ (reduce $\dfrac{3}{6}$ to $\dfrac{1}{2}$)

(b) $\dfrac{V}{IX} - \dfrac{I}{II} = \dfrac{V \times II}{IX \times II} - \dfrac{I \times IX}{II \times IX} = \dfrac{X}{XVIII} - \dfrac{IX}{XVIII} = \dfrac{X - IX}{XVIII} = \dfrac{I}{XVIII}$

Check: $\dfrac{5}{9} - \dfrac{1}{2} = \dfrac{5 \times 2}{9 \times 2} - \dfrac{1 \times 9}{2 \times 9} = \dfrac{10}{18} - \dfrac{9}{18} = \dfrac{10 - 9}{18} = \dfrac{1}{18}$

(c) $\dfrac{I}{V} + \dfrac{III}{IV} = \dfrac{I \times IV}{V \times IV} + \dfrac{III \times V}{IV \times V} = \dfrac{IV}{XX} + \dfrac{XV}{XX} = \dfrac{IV + XV}{XX} = \dfrac{XIX}{XX}$

Check: $\dfrac{1}{5} + \dfrac{3}{4} = \dfrac{1 \times 4}{5 \times 4} + \dfrac{3 \times 5}{4 \times 5} = \dfrac{4}{20} + \dfrac{15}{20} = \dfrac{4 + 15}{20} = \dfrac{19}{20}$

(d) $\dfrac{III}{IV} - \dfrac{II}{III} = \dfrac{III \times III}{IV \times III} - \dfrac{II \times IV}{III \times IV} = \dfrac{IX}{XII} - \dfrac{VIII}{XII} = \dfrac{IX-VIII}{XII} = \dfrac{I}{XII}$

Check: $\dfrac{3}{4} - \dfrac{2}{3} = \dfrac{3\times3}{4\times3} - \dfrac{2\times4}{3\times4} = \dfrac{9}{12} - \dfrac{8}{12} = \dfrac{9-8}{12} = \dfrac{1}{12}$

(e) $\dfrac{VII}{X} + \dfrac{IV}{XV} = \dfrac{VII \times III}{X \times III} + \dfrac{IV \times II}{XV \times II} = \dfrac{XXI}{XXX} + \dfrac{VIII}{XXX} = \dfrac{XXI+VIII}{XXX} = \dfrac{XXIX}{XXX}$

Check: $\dfrac{7}{10} + \dfrac{4}{15} = \dfrac{7\times3}{10\times3} + \dfrac{4\times2}{15\times2} = \dfrac{21}{30} + \dfrac{8}{30} = \dfrac{21+8}{30} = \dfrac{29}{30}$

(f) $\dfrac{XI}{XII} - \dfrac{V}{VIII} = \dfrac{XI \times II}{XII \times II} - \dfrac{V \times III}{VIII \times III} = \dfrac{XXII}{XXIV} - \dfrac{XV}{XXIV} = \dfrac{XXII-XV}{XXIV} = \dfrac{VII}{XXIV}$

Check: $\dfrac{11}{12} - \dfrac{5}{8} = \dfrac{11\times2}{12\times2} - \dfrac{5\times3}{8\times3} = \dfrac{22}{24} - \dfrac{15}{24} = \dfrac{22-15}{24} = \dfrac{7}{24}$

(g) $V - \dfrac{III}{IV} = \dfrac{V}{I} - \dfrac{III}{IV} = \dfrac{V \times IV}{I \times IV} - \dfrac{III \times I}{IV \times I} = \dfrac{XX}{IV} - \dfrac{III}{IV} = \dfrac{XX-III}{IV} = \dfrac{XVII}{IV}$ (alternatively, $IV\frac{I}{IV}$)

Check: $5 - \dfrac{3}{4} = \dfrac{5}{1} - \dfrac{3}{4} = \dfrac{5\times4}{1\times4} - \dfrac{3\times1}{4\times1} = \dfrac{20}{4} - \dfrac{3}{4} = \dfrac{20-3}{4} = \dfrac{17}{4}$ (alternatively, $4\frac{1}{4}$)

Page LII – Multiply Fractions

(a) $\dfrac{I}{II} \times \dfrac{V}{III} = \dfrac{I \times V}{II \times III} = \dfrac{V}{VI}$ $\left(\dfrac{1}{2} \times \dfrac{5}{3} = \dfrac{1\times5}{2\times3} = \dfrac{5}{6}\right)$

(b) $\dfrac{III}{IV} \times \dfrac{VIII}{IX} = \dfrac{III \times VIII}{IV \times IX} = \dfrac{XXIV}{XXXVI} = \dfrac{XXIV \div XII}{XXXVI \div XII} = \dfrac{II}{III}$ $\left(\dfrac{3}{4} \times \dfrac{8}{9} = \dfrac{3\times8}{4\times9} = \dfrac{24}{36} = \dfrac{24\div12}{36\div12} = \dfrac{2}{3}\right)$

Shortcut: $\dfrac{3}{4} \times \dfrac{8}{9} = \dfrac{3}{9} \times \dfrac{8}{4} = \dfrac{1}{3} \times \dfrac{2}{1} = \dfrac{2}{3}$

(c) $\dfrac{V}{VIII} \times \dfrac{IV}{XXV} = \dfrac{V \times IV}{VIII \times XXV} = \dfrac{XX}{CC} = \dfrac{XX \div XX}{CC \div XX} = \dfrac{I}{X}$ $\left(\dfrac{5}{8} \times \dfrac{4}{25} = \dfrac{5\times4}{8\times25} = \dfrac{20}{200} = \dfrac{20\div20}{200\div20} = \dfrac{1}{10}\right)$

Shortcut: $\dfrac{5}{8} \times \dfrac{4}{25} = \dfrac{5}{25} \times \dfrac{4}{8} = \dfrac{1}{5} \times \dfrac{1}{2} = \dfrac{1}{10}$

(d) $\dfrac{VII}{X} \times \dfrac{L}{XLIX} = \dfrac{VII \times L}{X \times XLIX} = \dfrac{CCCL}{CDXC} = \dfrac{CCCL \div LXX}{CDXC \div LXX} = \dfrac{V}{VII}$ $\left(\dfrac{7}{10} \times \dfrac{50}{49} = \dfrac{7\times50}{10\times49} = \dfrac{350}{490} = \dfrac{350\div70}{490\div70} = \dfrac{5}{7}\right)$

Shortcut: $\dfrac{7}{10} \times \dfrac{50}{49} = \dfrac{7}{49} \times \dfrac{50}{10} = \dfrac{1}{7} \times \dfrac{5}{1} = \dfrac{5}{7}$

(e) $\dfrac{VI}{LV} \times \dfrac{XXII}{III} = \dfrac{VI \times XXII}{LV \times III} = \dfrac{CXXXII}{CLXV} = \dfrac{CXXXII \div XXXIII}{CLXV \div XXXIII} = \dfrac{IV}{V}$ $\left(\dfrac{6}{55} \times \dfrac{22}{3} = \dfrac{6\times22}{55\times3} = \dfrac{132}{165} = \dfrac{132\div33}{165\div33} = \dfrac{4}{5}\right)$

Shortcut: $\dfrac{6}{55} \times \dfrac{22}{3} = \dfrac{6}{3} \times \dfrac{22}{55} = \dfrac{2}{1} \times \dfrac{2}{5} = \dfrac{4}{5}$

(f) $\dfrac{VII}{L} \times \dfrac{C}{XXI} = \dfrac{VII \times C}{L \times XXI} = \dfrac{DCC}{ML} = \dfrac{DCC \div CCCL}{ML \div CCCL} = \dfrac{II}{III}$ $\left(\dfrac{7}{50} \times \dfrac{100}{21} = \dfrac{7\times100}{5\times21} = \dfrac{700}{1050} = \dfrac{700\div350}{1050\div350} = \dfrac{2}{3}\right)$

Shortcut: $\dfrac{7}{50} \times \dfrac{100}{21} = \dfrac{7}{21} \times \dfrac{100}{50} = \dfrac{1}{3} \times \dfrac{2}{1} = \dfrac{2}{3}$

(g) $\dfrac{III}{M} \times \dfrac{CD}{IX} = \dfrac{III \times CD}{M \times IX} = \dfrac{MCC}{M \times IX} = \dfrac{MCC \div C}{M \times IX \div C} = \dfrac{XII}{X \times IX} = \dfrac{XII}{XC} = \dfrac{XII \div VI}{XC \div VI} = \dfrac{II}{XV}$

$\left(\dfrac{3}{1000} \times \dfrac{400}{9} = \dfrac{3\times400}{1000\times9} = \dfrac{1200}{9000} = \dfrac{1200\div600}{9000\div600} = \dfrac{2}{15}\right)$

Shortcut (and much simpler method): $\dfrac{3}{1000} \times \dfrac{400}{9} = \dfrac{3}{9} \times \dfrac{400}{1000} = \dfrac{1}{3} \times \dfrac{2}{5} = \dfrac{2}{15}$

Page LIII – Reciprocals

(a) The reciprocal of $\frac{V}{VI}$ equals $\frac{VI}{V}$. (Alternate answer: $I\frac{I}{V}$)

(b) The reciprocal of $\frac{III}{VIII}$ equals $\frac{VIII}{III}$. (Alternate answer: $II\frac{II}{III}$)

(c) The reciprocal of V equals $\frac{I}{V}$.

(d) The reciprocal of $\frac{IX}{IV}$ equals $\frac{IV}{IX}$.

(e) The reciprocal of $\frac{V}{II}$ equals $\frac{II}{V}$.

(f) The reciprocal of $\frac{I}{L}$ equals L. Note that $\frac{L}{1} = L$.

(g) The reciprocal of $\frac{XI}{XC}$ equals $\frac{XC}{XI}$. (Alternate answer: $VIII\frac{II}{XI}$)

Page LIV – Divide Fractions

(a) $\frac{I}{II} \div \frac{III}{IV} = \frac{I}{II} \times \frac{IV}{III} = \frac{I \times IV}{II \times III} = \frac{IV}{VI} = \frac{IV \div II}{VI \div II} = \frac{II}{III}$ $\left(\frac{1}{2} \div \frac{3}{4} = \frac{1}{2} \times \frac{4}{3} = \frac{1 \times 4}{2 \times 3} = \frac{4}{6} = \frac{4 \div 2}{6 \div 2} = \frac{2}{3}\right)$

(b) $\frac{II}{III} \div \frac{IV}{IX} = \frac{II}{III} \times \frac{IX}{IV} = \frac{II \times IX}{III \times IV} = \frac{XVIII}{XII} = \frac{XVIII \div VI}{XII \div VI} = \frac{III}{II}$ (Alternate answer: $I\frac{I}{II}$)

$\left(\frac{2}{3} \div \frac{4}{9} = \frac{2}{3} \times \frac{9}{4} = \frac{2 \times 9}{3 \times 4} = \frac{18}{12} = \frac{18 \div 6}{12 \div 6} = \frac{3}{2} \text{ or } 1\frac{1}{2}\right)$

(c) $\frac{III}{X} \div \frac{VII}{V} = \frac{III}{X} \times \frac{V}{VII} = \frac{III \times V}{X \times VII} = \frac{XV}{LXX} = \frac{XV \div V}{LXX \div V} = \frac{III}{XIV}$

$\left(\frac{3}{10} \div \frac{7}{5} = \frac{3}{10} \times \frac{5}{7} = \frac{3 \times 5}{10 \times 7} = \frac{15}{70} = \frac{15 \div 5}{70 \div 5} = \frac{3}{14}\right)$ Short: $\frac{3}{10} \times \frac{5}{7} = \frac{3}{7} \times \frac{5}{10} = \frac{3}{7} \times \frac{1}{2} = \frac{3}{14}$

(d) $\frac{IV}{VII} \div \frac{V}{II} = \frac{IV}{VII} \times \frac{II}{V} = \frac{IV \times II}{VII \times V} = \frac{VIII}{XXXV}$ $\left(\frac{4}{7} \div \frac{5}{2} = \frac{4}{7} \times \frac{2}{5} = \frac{4 \times 2}{7 \times 5} = \frac{8}{35}\right)$

(e) $\frac{V}{VI} \div \frac{I}{III} = \frac{V}{VI} \times \frac{III}{I} = \frac{V \times III}{VI \times I} = \frac{XV}{VI} = \frac{XV \div III}{VI \div III} = \frac{V}{II}$ (Alternate answer: $II\frac{I}{II}$)

$\left(\frac{5}{6} \div \frac{1}{3} = \frac{5}{6} \times \frac{3}{1} = \frac{5 \times 3}{6 \times 1} = \frac{15}{6} = \frac{15 \div 3}{6 \div 3} = \frac{5}{2} \text{ or } 2\frac{1}{2}\right)$

(f) $\frac{III}{L} \div \frac{I}{C} = \frac{III}{L} \times \frac{C}{I} = \frac{III \times C}{L \times I} = \frac{CCC}{L} = \frac{CCC \div L}{L \div L} = \frac{VI}{I} = VI$

$\left(\frac{3}{50} \div \frac{1}{100} = \frac{3}{50} \times \frac{100}{1} = \frac{3 \times 100}{50 \times 1} = \frac{300}{50} = \frac{300 \div 50}{50 \div 50} = \frac{6}{1} = 6 \text{ or } \frac{300}{50} = 300 \div 50 = 6\right)$

(g) $\frac{IX}{D} \div \frac{XI}{M} = \frac{IX}{D} \times \frac{M}{XI} = \frac{IX \times M}{D \times XI} = \frac{IX}{XI} \times \frac{M}{D} = \frac{IX}{XI} \times \frac{II}{I} = \frac{IX \times II}{XI \times I} = \frac{XVIII}{XI}$ (Alternate answer: $I\frac{VII}{XI}$)

$\left(\frac{9}{500} \div \frac{11}{1000} = \frac{9}{500} \times \frac{1000}{11} = \frac{9 \times 1000}{500 \times 11} = \frac{9}{11} \times \frac{1000}{500} = \frac{9}{11} \times \frac{2}{1} = \frac{18}{11} \text{ or } 1\frac{7}{11}\right)$

Page LV – Percents

(a) $L\% = \frac{L}{C} = \frac{L \div L}{C \div L} = \frac{I}{II} \left(50\% = \frac{50}{100} = \frac{50 \div 50}{100 \div 50} = \frac{1}{2}\right)$

(b) $XXV\% = \frac{XXV}{C} = \frac{XXV \div XXV}{C \div XXV} = \frac{I}{IV} \left(25\% = \frac{25}{100} = \frac{25 \div 25}{100 \div 25} = \frac{1}{4}\right)$

(c) $LXXX\% = \frac{LXXX}{C} = \frac{LXXX \div XX}{C \div XX} = \frac{IV}{V} \left(80\% = \frac{80}{100} = \frac{80 \div 20}{100 \div 20} = \frac{4}{5}\right)$

(d) $CL\% = \frac{CL}{C} = \frac{CL \div L}{C \div L} = \frac{III}{II}$ or $I\frac{I}{II} \left(150\% = \frac{150}{100} = \frac{150 \div 50}{100 \div 50} = \frac{3}{2}$ or $1\frac{1}{2}\right)$

(e) $CD\% = \frac{CD}{C} = \frac{CD \div C}{C \div C} = \frac{IV}{I} = IV \left(400\% = \frac{400}{100} = \frac{400 \div 100}{100 \div 100} = \frac{4}{1} = 4\right)$

(f) $XV\% = \frac{XV}{C} = \frac{XV \div V}{C \div V} = \frac{III}{XX} \left(15\% = \frac{15}{100} = \frac{15 \div 5}{100 \div 5} = \frac{3}{20}\right)$

(g) $LXIV\% = \frac{LXIV}{C} = \frac{LXIV \div IV}{C \div IV} = \frac{XVI}{XXV} \left(64\% = \frac{64}{100} = \frac{64 \div 4}{100 \div 4} = \frac{16}{25}\right)$

VII – Number Pattern Puzzles

Page LVIII – Set I (Easy)

(a) XLI, XLIX (add 8)

$9 + 8 = 17, 17 + 8 = 25, 25 + 8 = 33, 33 + 8 = 41, 41 + 8 = 49$

(b) XVI, IV (subtract 12)

$64 - 12 = 52, 52 - 12 = 40, 40 - 12 = 28, 28 - 12 = 16, 16 - 12 = 4$

(c) XCI, CXII (add 17, add 18, add 19, add 20, add 21)

$17 + 17 = 34, 34 + 18 = 52, 52 + 19 = 71, 71 + 20 = 91, 91 + 21 = 112$

(d) XLIV, XIX (subtract 25)

$144 - 25 = 119, 119 - 25 = 94, 94 - 25 = 69, 69 - 25 = 44, 44 - 25 = 19$

(e) XXXVIII, LI (add 5, add 7, add 9, add 11, add 13)

$6 + 5 = 11, 11 + 7 = 18, 18 + 9 = 27, 27 + 11 = 38, 38 + 13 = 51$

(f) CVI, XCIV (subtract 60, subtract 48, subtract 36, subtract 24, subtract 12)

$274 - 60 = 214, 214 - 48 = 166, 166 - 36 = 130, 130 - 24 = 106, 106 - 12 = 94$

(g) XII, III (divide by 4)

$3072 \div 4 = 768, 768 \div 4 = 192, 192 \div 4 = 48, 48 \div 4 = 12, 12 \div 4 = 3$

(h) D, M (these are the Roman numerals that consist of a single letter only)

(i) CDLXIV, XLI (subtract 423): $2156 - 423 = 1733, 1733 - 423 = 1310,$
$1310 - 423 = 887, 887 - 423 = 464, 464 - 423 = 41$

(j) CXX, DCCXX (multiply by 2, multiply by 3, multiply by 4, multiply by 5, etc.)

$1 \times 2 = 2, 2 \times 3 = 6, 6 \times 4 = 24, 24 \times 5 = 120, 120 \times 6 = 720$

Page LIX – Set II (Easy)

(a) LXXXI, C (5 squared, 6 squared, 7 squared, 8 squared, 9 squared, etc.)

$5^2 = 25, 6^2 = 36, 7^2 = 49, 8^2 = 64, 9^2 = 81, 10^2 = 100$

(b) XXXVII, LX (add the previous two numbers, like the Fibonacci sequence)

$4, 5, 4 + 5 = 9, 5 + 9 = 14, 9 + 14 = 23, 14 + 23 = 37, 23 + 37 = 60$

(c) LXIV, LXXXV (add 9, add 12, add 15, add 18, add 21)

$10 + 9 = 19, 19 + 12 = 31, 31 + 15 = 46, 46 + 18 = 64, 64 + 21 = 85$

(d) MDC, MMMCC (multiply by 2): $100 \times 2 = 200, 200 \times 2 = 400,$

$400 \times 2 = 800, 800 \times 2 = 1600, 1600 \times 2 = 3200$

(e) XIX, IV (subtract the two previous numbers; Fibonacci-inspired)

$172 - 107 = 65, 107 - 65 = 42, 65 - 42 = 23, 42 - 23 = 19, 23 - 19 = 4$

(f) CXXX, CCLVIII (add 2, add 4, add 8, add 16, add 32, add 64, add 128)

$4 + 2 = 6, 6 + 4 = 10, 10 + 8 = 18, 18 + 16 = 34, 34 + 32 = 66,$

$66 + 64 = 130, 130 + 128 = 258$

(g) VI, I (divide by 2, divide by 3, divide by 4, divide by 5, divide by 6)

$720 \div 2 = 360, 360 \div 3 = 120, 120 \div 4 = 30, 30 \div 5 = 6, 6 \div 6 = 1$

(h) LXIV, XXVII (8 cubed, 7 cubed, 6 cubed, 5 cubed, 4 cubed, 3 cubed)

$8^3 = 512, 7^3 = 343, 6^3 = 216, 5^3 = 125, 4^3 = 64, 3^3 = 27$

(i) CXII, LVI (divide by 2)

$1792 \div 2 = 896, 896 \div 2 = 448, 448 \div 2 = 224, 224 \div 2 = 112, 112 \div 2 = 56$

(j) XLI, XLIII (these are prime numbers)

$17, 19, 23, 29, 31, 37, 41, 43$

Page LX – Set III (Medium)

(a) CCCLXXVI, DCCLX (multiply by 2 and add 8)

$4 \times 2 + 8 = 16, 16 \times 2 + 8 = 40, 40 \times 2 + 8 = 88, 88 \times 2 + 8 = 184,$

$184 \times 2 + 8 = 376, 376 \times 2 + 8 = 760$

(b) CXXV, CLX (there are two sequences merged together: one sequence multiplies every other number by 2, while the other sequence divides every other number by 2)

$5 \times 2 = 10, 10 \times 2 = 20, 20 \times 2 = 40, 40 \times 2 = 80, 80 \times 2 = 160$

$2000 \div 2 = 1000, 1000 \div 2 = 500, 500 \div 2 = 250, 250 \div 2 = 125$

5, 2000, 10, 1000, 20, 500, 40, 250, 80, 125, 160

(c) CCXXX, CDLX (multiply by 2, add 10, multiply by 2, add 10, etc.)

$5 \times 2 = 10, 10 + 10 = 20, 20 \times 2 = 40, 40 + 10 = 50, 50 \times 2 = 100,$

$100 + 10 = 110, 110 \times 2 = 220, 220 + 10 = 230, 230 \times 2 = 460$

(d) LIX, LXVII (skip every other prime number)

2 (skip 3) 5 (skip 7) 11 (skip 13) 17 (skip 19) 23 (skip 29) 31 (skip 37) 41 (skip 43) 47 (skip 53) 59 (skip 61) 67

(e) CXXVIII, LXIV (multiply by 4, divide by 2, multiply by 4, divide by 2, etc.)

$4 \times 4 = 16, 16 \div 2 = 8, 8 \times 4 = 32, 32 \div 2 = 16, 16 \times 4 = 64, 64 \div 2 = 32,$

$32 \times 4 = 128, 128 \div 2 = 64$

(f) CCCXX, LX (there are two sequences merged together: one sequence multiplies every other number by 2, while the other sequence subtracts 10 from every other number)

$10 \times 2 = 20, 20 \times 2 = 40, 40 \times 2 = 80, 80 \times 2 = 160, 160 \times 2 = 320$

$110 - 10 = 100, 100 - 10 = 90, 90 - 10 = 80, 80 - 10 = 70, 70 - 10 = 60$

10, 110, 20, 100, 40, 90, 80, 80, 160, 70, 320, 60

(g) LD, XD (these are Roman numerals consisting of two letters which are made from irregular subtractions; see the last sections of Chapters 1 and 2)

VX, VL, IL, LC, VC, IC, LD, XD

(h) LXI, LXV (these are Roman numerals which consist of exactly three different letters): XIV, XVI, XLI, XLV, LIV, LVI, LIX, LXI, LXV

(i) CCLIII, MMXXIV (add 1, times 2, add 3, times 4, add 5, times 6, add 7, etc.)

$2 + 1 = 3, 3 \times 2 = 6, 6 + 3 = 9, 9 \times 4 = 36, 36 + 5 = 41, 41 \times 6 = 246,$

$246 + 7 = 253, 253 \times 8 = 2024$

(j) XXII, XLVII (there are two sequences merged together: one sequence adds 8 to every other number, while the other sequence subtracts 7 from every other number)

$7 + 8 = 15, 15 + 8 = 23, 23 + 8 = 31, 31 + 8 = 39, 39 + 8 = 47$

$50 - 7 = 43, 43 - 7 = 36, 36 - 7 = 29, 29 - 7 = 22$

7, 50, 15, 43, 23, 36, 31, 29, 39, 22, 47

Page LXI – Set IV (Hard)

    (a) XXV, XXX (these are Roman numerals consisting of exactly three letters)

III, VII, XII, XIV, XVI, XIX, XXI, XXV, XXX

    (b) CCCXLV, DXIV (2 cubed plus 2, 3 cubed plus 2, 4 cubed plus 2, etc.)

$2^3 + 2 = 10, 3^3 + 2 = 29, 4^3 + 2 = 66, 5^3 + 2 = 127, 6^3 + 2 = 218,$

$7^3 + 2 = 345, 8^3 + 2 = 514$

    (c) CXLIX, CCLXXIV (add the previous three numbers; Fibonacci-inspired)

$1 + 2 + 4 = 7, 2 + 4 + 7 = 13, 4 + 7 + 13 = 24, 7 + 13 + 24 = 44,$

$13 + 24 + 44 = 81, 24 + 44 + 81 = 149, 44 + 81 + 149 = 274$

    (d) LVIII, LXII (multiply the prime numbers by 2)

$2 \times 2 = 4, 3 \times 2 = 6, 5 \times 2 = 10, 7 \times 2 = 14, 11 \times 2 = 22, 13 \times 2 = 26,$

$17 \times 2 = 34, 19 \times 2 = 38, 23 \times 2 = 46, 29 \times 2 = 58, 31 \times 2 = 62$

    (e) CCC, MCM, (these are the regular Roman numerals consisting of two or more letters that are the same when the letters are reversed)

II, III, XIX, XX, XXX, CXC, CC, CCC, MCM

    (f) CCLXXVII, CDLIV (add the first two numbers plus 1, add the second two numbers plus 2, add the next two numbers plus 3, add the next two numbers plus 4, add the next two numbers plus 5, etc.; Fibonacci-inspired)

$(2 + 3) + 1 = 6, (3 + 6) + 2 = 11, (6 + 11) + 3 = 20, (11 + 20) + 4 = 35,$

$(20 + 35) + 5 = 60, (35 + 60) + 6 = 101, (60 + 101) + 7 = 168,$

$(101 + 168) + 8 = 277, (168 + 277) + 9 = 454$

    (g) XXX, XXXII (add one to each prime number)

$2 + 1 = 3, 3 + 1 = 4, 5 + 1 = 6, 7 + 1 = 8, 11 + 1 = 12, 13 + 1 = 14,$

$17 + 1 = 18, 19 + 1 = 20, 23 + 1 = 24, 29 + 1 = 30, 31 + 1 = 32$

    (h) DCCXXIX, CXXVIII (8 to the power of 1, 7 to the power of 2, 6 to the power of 3, 5 to the power of 4, 4 to the power of 5, 3 to the power of 6, 2 to the power of 7)

$8^1 = 8, 7^2 = 49, 6^3 = 216, 5^4 = 625, 4^5 = 1024, 3^6 = 729, 2^7 = 128$

(i) XXIX, XXIX (there are two sequences merged together: one sequence adds 5 to every other number, while the other sequence adds 4 to every other number)

$4 + 5 = 9, 9 + 5 = 14, 14 + 5 = 19, 19 + 5 = 24, 24 + 5 = 29$

$9 + 4 = 13, 13 + 4 = 17, 17 + 4 = 21, 21 + 4 = 25, 25 + 4 = 29$

4, $\boxed{9}$, 9, $\boxed{13}$, 14, $\boxed{17}$, 19, $\boxed{21}$, 24, $\boxed{25}$, 29, $\boxed{29}$

(j) CMLXI, MCCCLXIX (square prime numbers starting with 11)

$11^2 = 121, 13^2 = 169, 17^2 = 289, 19^2 = 361, 23^2 = 529, 29^2 = 841,$

$31^2 = 961, 37^2 = 1369$

Page LXII – Set V (II×II Arrays)

(a) XII (add the left squares, then divide by the top right square)

 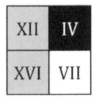

$(9 + 15) \div 3 = 8, (12 + 16) \div 4 = 7, (50 + 100) \div 10 = 15, (100 + 500) \div 50 = 12$

(b) CC (subtract the top right from the bottom left, then multiply by the top left)

 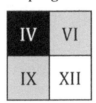 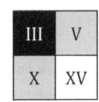 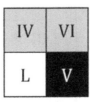

$(5 - 3) \times 2 = 4, (9 - 6) \times 4 = 12, (10 - 5) \times 3 = 15, (50 - 10) \times 5 = 200$

(c) LX (rotating puzzle: add the gray squares, then multiply by the black square)

 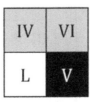 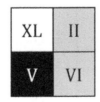 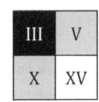

$(3 + 2) \times 4 = 20, (4 + 6) \times 5 = 50, (2 + 6) \times 5 = 40, (1 + 5) \times 10 = 60$

(d) CL (multiply the top right by the bottom left, then subtract the top left)

V	II
III	I

VI	III
IV	VI

IX	IV
VI	XV

C	V
L	CL

$2 \times 3 - 5 = 1, 3 \times 4 - 6 = 6, 4 \times 6 - 9 = 15, 5 \times 50 - 100 = 150$

Page LXIII – Set VI (III×III Arrays)

(a) Add the last two numbers (Fibonacci-inspired): $2 + 4 = 6, 4 + 6 = 10$, $6 + 10 = 16, 10 + 16 = 26, 16 + 26 = 42, 26 + 42 = 68, 42 + 68 = 110$

(b) Square even numbers: $2^2 = 4, 4^2 = 16, 6^2 = 36, 8^2 = 64, 10^2 = 100$, $12^2 = 144, 14^2 = 196, 16^2 = 256, 18^2 = 324$

II	IV	VI
XXVI	XVI	X
XLII	LXVIII	CX

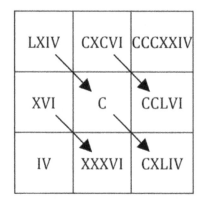

LXIV	CXCVI	CCCXXIV
XVI	C	CCLVI
IV	XXXVI	CXLIV

(c) These are the Roman numerals consisting of exactly four letters which involve two regular subtractions.

(d) $1 + 1 = 2, 2 + 2 = 4, 4 + 3 = 7, 7 + 4 = 11, 11 + 5 = 16, 16 + 6 = 22$, $22 + 7 = 29, 29 + 8 = 37$

CMIV	CDIX	XCIV
CDXC	CDIV	XLIX
CDXL	XCIX	XLIV

XXII	XVI	I
XXIX	XI	II
XXXVII	VII	IV

Page LXIV – Set VII (I+II Pyramids)

(a) LIV (multiply the left square by the top square)

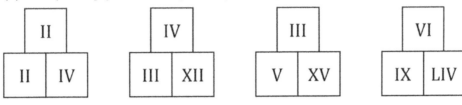

$2 \times 2 = 4, 3 \times 4 = 12, 5 \times 3 = 15, 9 \times 6 = 54$

(b) V (rotating puzzle: subtract the gray square from the black square)

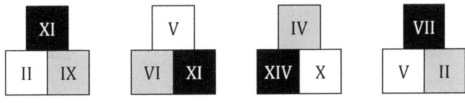

$11 - 9 = 2, 11 - 6 = 5, 14 - 4 = 10, 7 - 2 = 5$

(c) MDC (subtract the right square from the top square, but then square this value to get the left square)

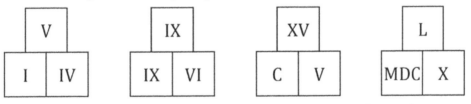

$(5 - 4)^2 = 1, (9 - 6)^2 = 9, (15 - 5)^2 = 100, (50 - 10)^2 = 40^2 = 1600$

(d) LXV (multiply the bottom squares, then add the bottom squares, and then add these two values – the product and the sum – together)

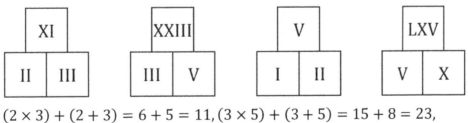

$(2 \times 3) + (2 + 3) = 6 + 5 = 11, (3 \times 5) + (3 + 5) = 15 + 8 = 23,$

$(1 \times 2) + (1 + 2) = 2 + 3 = 5, (5 \times 10) + (5 + 10) = 50 + 15 = 65$

Page LXV – Set VIII (I+II+III Pyramids)

(a) Multiply the numbers below each square: $2 \times 5 = 10, 5 \times 3 = 15,$
$10 \times 15 = 150$

(b) Raise numbers to the fourth power, going down to the right:
$2^4 = 16, 3^4 = 81, 4^4 = 256, 5^4 = 625, 6^4 = 1296, 7^4 = 2401$

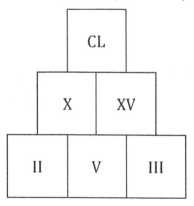
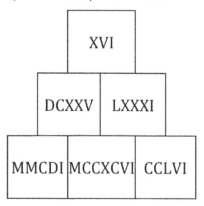

(c) These are Roman numerals consisting of exactly two letters, arranged in order along a clockwise path.

(d) First add the numbers below and then square this value.
$(1 + 2)^2 = 3^2 = 9, (2 + 4)^2 = 6^2 = 36, (9 + 36)^2 = 45^2 = 2025$

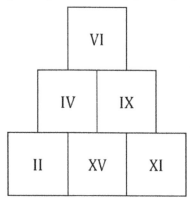
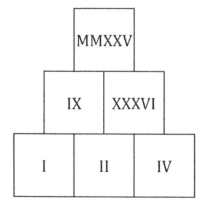

Page LXVI – Set IX (I+II+III+IV+V Pyramid)

Going up to the left along the top right diagonal, square integer values:

$2^2 = 4, 3^2 = 9, 4^2 = 16, 5^2 = 25, 6^2 = 36$

One way to find the other values is:

Double the value above and subtract the value to the right.

$2 \times 9 - 4 = 14, 2 \times 16 - 9 = 23, 2 \times 25 - 16 = 34, 2 \times 36 - 25 = 47$

$2 \times 23 - 14 = 32, 2 \times 34 - 23 = 45, 2 \times 47 - 34 = 60$

$2 \times 45 - 32 = 58, 2 \times 60 - 45 = 75$

$2 \times 75 - 58 = 92$

Alternate method: Find the pattern of how much to add going left.

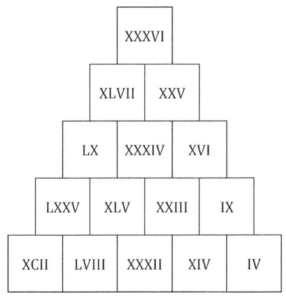

VIII – Variety Puzzles

Page LXVIII – Roman Numeral Builder

27 Roman numerals: I, II, IV, V, VI, VII, L, LI, LII, LIV, LV, LVI, LVII, C, CI, CII, CIV, CV, CVI, CVII, CL, CLI, CLII, CLIV, CLV, CLVI, CLVII

Page LXIX – Word Builder

Our 28 answers (based on which words are allowed in Scrabble) include: CIVIC, CIVIL, DID, DILL, DIM, ILL, IMID, IMIDIC, IMMIX, LI, LID, LIVID, MI, MID, MIDI, MIL, MILD, MILL, MIM, MIMIC, MIX, MM, VILL, VILLI, VIM, VIVID, XI, ID

Page LXX – Roman Numeral Fill-in

13	XIII	85	LXXXV	28	XXVIII
350	CCCL	169	CLXIX	249	CCXLIX
750	DCCL	175	CLXXV	1085	MLXXXV
2600	MMDC	1705	MDCCV	1850	MDCCCL

38	XXXVIII	88	LXXXVIII
87	LXXXVII	286	CCLXXXVI
685	DCLXXXV	1676	MDCLXXVI
1975	MCMLXXV	2980	MMCMLXXX

M				M		C	L	X
D	C	C	L	M	D	C	L	X	X	V	I
C				D		X		X		I
C		M	M	C	M	L	X	X	X		I
C	C	C	L		D		V		V
L			X		C			I
	M		X		C	C	X	L	I	X
D	C	L	X	X	X	V				X
	M		V			C		L		X
	L			C	C	L	X	X	X	V	I
	X				X		X		I
L	X	X	X	V	I	I	I		X		I
	V				X	X	V	I	I	I

Page LXXI – Cryptograms

(a) The colosseum had underground passages and trap doors that could be used for elaborate special effects.

```
E T A B V R Z K W I S U L F G O H P Q Y D N M C X J
A B C D E F G H I J K L M N O P Q R S T U V W X Y Z
```

(b) Pompeii was a popular resort for wealthy Romans until it was destroyed by the volcano, Mount Vesuvius.

```
W G N J B S Z K O Y R T H Q I D P A L C X M F V E U
A B C D E F G H I J K L M N O P Q R S T U V W X Y Z
```

Page LXXII – Roman Numeral Sudoku

(a)

IV	II	V	I	III	VI
III	V	VI	IV	I	II
VI	I	III	II	V	IV
II	IV	I	V	VI	III
V	III	II	VI	IV	I
I	VI	IV	III	II	V

(b)

VI	I	II	IV	III	V
III	IV	V	I	VI	II
II	V	VI	III	I	IV
IV	III	I	V	II	VI
I	II	IV	VI	V	III
V	VI	III	II	IV	I

(c)

IV	II	VI	I	V	III
VI	I	III	V	II	IV
III	V	II	IV	VI	I
II	IV	I	VI	III	V
V	III	IV	II	I	VI
I	VI	V	III	IV	II

(d)

III	V	I	IV	II	VI
II	IV	V	VI	III	I
VI	I	III	II	V	IV
I	VI	II	III	IV	V
IV	II	VI	V	I	III
V	III	IV	I	VI	II

Page LXXIII – Roman Numeral Kakuro

First, examine the easiest answers, which are filled in below.
(For the completed puzzle, see the following page.)

	XVII	XXI				III	VI	XVI	XIV			XXI	XI
XVI	IX	VII	IV	XII	XIII			VII			XII / IV		
XV	VIII				XVI			IX		VII / XXX	I		
VI / X					IX	XIX		XX / XII			III		
IV	I	III	XXIII				XXIV / XIV	IX			III / XX		
XVII	IX	VIII		XI				V	III	XXII			
			XVII	XVI	XIII / III	IV	IX			XIV			IV
		XIX									IV	I	III
		XXI									III	II	I

On the top left, 16 must be 7 and 9, while 17 must be 8 and 9. (Note that 16 can't be 8 and 8 because digits can't be repeated.) Similarly, on the middle left, 17 must be 9 and 8, and only the 9 can be part of the 10 (since if the 8 is part of the 10, you would need 2 and 2 to make the 4, but you aren't allowed to repeat 2's).

Near the top middle, 16 must be 9 and 7, but only the 7 is small enough to be included with the 13.

In the middle, 11 must be 1, 2, 3, and 5, but only the 5 is big enough to be part of the 14, and then only the 3 can be part of the 12.

In the bottom right, 4 must be 1 and 3, while 3 must be 1 and 2.

On the top right, 7 must be 1, 2, and 4, so the 3 that makes the 4 can't be part of the 7.

The completed Kakuro puzzle is shown below.

	XVII	XXI			III	VI	XVI	XIV			XXI	XI
XVI	IX	VII	IV / XII		XIII / I	I	II	VII	III		XII / IV	IX	III
XV	VIII	II	I	IV	XVI	II	IV	IX	I	VII / XXX	I	IV	II
VI / X		I	III	II	IX	XIX	XX / XII	II	IX	III	I	V
IV	I	III	XXIII	VI	VIII	IX	XXIV / XIV	IX	VIII	VII	III / XX	II	I
XVII	IX	VIII		XI	I	II	V	III	XXII	VIII	IX	V
			XVII	XVI	XIII / III	IV	IX		XIV	VI	VIII	IV
	XIX	IX	VII	II	I				IV	I	III
	XXI	VIII	IX	I	III				III	II	I

A little trial and error may help with the bottom left. Note that the 17 must be 9 and 8, while the 16 must be 9 and 7, and the 3 must be 1 and 2.

Near the right middle, note that 24 must be 9, 8, and 7, while 30 must be 9, 8, 7, and 6.

At the top middle, once the 7 and 9 are put in, the remaining numbers of the 13 must be 1, 2, and 3, while the remaining numbers of the 16 must be 1, 2, and 4.

Some trial and error may help with the top right, but you shouldn't need too much trial and error if you apply enough logic and reasoning.

Page LXXIV – Word Scrambles

  (a) AQUEDUCT   (b) ABACUS

  (c) EMPEROR   (d) STYLUS

  (e) COHORT   (f) RHETORIC

  (g) CENTURY   (h) CITIZEN

  (i) bonus: REPUBLIC

Page LXXV – Word Search

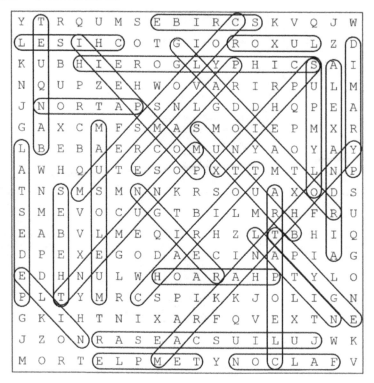

Page LXXVI – Roman Numeral Math Crossword

Below you can find the answers to the math crossword.

On the next page, you can check how we arrived at our answers.

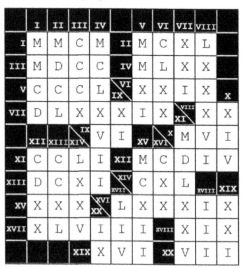

ACROSS

I. MMM − C = 3000 − 100 = 2900 = MMCM

II. CXC × VI = 190 × 6 = 1140 = MCXL

III. MM − CCC = 2000 − 300 = 1700 = MDCC

IV. MC − XXX = 1100 − 30 = 1070 = MLXX

V. ML ÷ III = 1050 ÷ 3 = 350 = CCCL

VI. XXX − I = 30 − 1 = 29 = XXIX

VII. DC − XI = 600 − 11 = 589 = DLXXXIX

VIII. $\sqrt{CD} = \sqrt{400} = 20 = XX$

IX. I + II + III = 1 + 2 + 3 = 6 = VI

X. DIII × II = 503 × 2 = 1006 = MVI

XI. MIV ÷ IV = 1004 ÷ 4 = 251 = CCLI

XII. MD − XCVI = 1500 − 96 = 1404 = MCDIV

XIII. DX + CI = 510 + 101 = 611 = DCXI

XIV. XII^{II} − IV = 12^2 − 4 = 144 − 4 = 140 = CXL

XV. V × VI = 5 × 6 = 30 = XXX

XVI. C − XI = 100 − 11 = 89 = LXXXIX

XVII. L − II = 50 − 2 = 48 = XLVIII

XVIII. XX − I = 20 − 1 = 19 = XIX

XIX. $II^{IV} = 2^4$ = 2 × 2 × 2 × 2 = 4 × 4 = 16 = XVI

XX. $\sqrt{XLIX} = \sqrt{49}$ = 7 = VII

DOWN

I. L^{II} − C = 50^2 − 100 = 2500 − 100 = 2400 = MMCD

II. CL × XI = 150 × 11 = 1650 = MDCL

III. CD − XC = 400 − 90 = 310 = CCCX

IV. MCC − XXXIII = 1200 − 33 = 1167 = MCLXVII

V. MX + MI = 1010 + 1001 = 2011 = MMXI

VI. CC − XXX = 200 − 30 = 170 = CLXX

VII. III × VII = 3 × 7 = 21 = XXI

VIII. C − XIV = 100 − 14 = 86 = LXXXVI

IX. $XX - IX = 20 - 9 = 11 = XI$

X. $II \times VII = 2 \times 7 = 14 = XIV$

XI. $XV \times CV = 15 \times 105 = 1575 = MDLXXV$

XII. $D - LXXX = 500 - 80 = 420 = CDXX$

XIII. $VI \times XL = 6 \times 40 = 240 = CCXL$

XIV. $CL \div II = 150 \div 2 = 75 = LXXV$

XV. $MIX + CIII = 1009 + 103 = 1112 = MCXII$

XVI. $III \times XL = 3 \times 40 = 120 = CXX$

XVII. $VI \times IX = 6 \times 9 = 54 = LIV$

XVIII. $\sqrt{IX} = \sqrt{9} = 3 = III$

XIX. $CV \div V = 105 \div 5 = 21 = XXI$

XX. $\sqrt{LXXXI} = \sqrt{81} = 9 = IX$

IX – Word Problems

Page LXXVIII – Age Problems

(a) LXXV (75) years (note that 76 and 77 are incorrect, but common)
There are two "tricks" to this problem.
If you did $14 - (-63) = 14 + 63 = 77$, you have the right idea, but you need to make a couple of corrections in order to get the right answer. First of all, there is no year "zero," which reduces 77 down to 76. (For example, a person born on September 23 in 1 BC would celebrate his or her first birthday on September 23, 1 AD, since there is no 0 in between.)
Second of all, Augustus was born in September, but died in August. If he had died on September 23 instead, then he would have reached his 76th birthday. But since he died in August, he was still 75 at that time.
(b) Atticus died in LXXXVIII (88) BC and Cecilia died in LXXXV (85) BC.
For Atticus, $147 - 59 = 88$, and three months after his 59th birthday in May, it was still 88 BC. He died in August of 88 BC.
For Cecilia, $147 - 61 = 86$, but 8 months after her 61st birthday in May, it was already 85 BC (one less than 86). She died in January of 85 BC.

Page LXXIX – Comparison Problems

(a) Felix had LXXII (72) grapes and Cyrus had XXIV (24) grapes.

Check: There are a total of $72 + 24 = 96$ grapes.

Originally, Felix had $\frac{96}{2} = 48$ grapes and Cyrus also had 48 grapes.

When Cyrus gave 24 grapes to Felix:

Felix had $48 + 24 = 72$ grapes.

Cyrus had $48 - 24 = 24$ grapes.

Since $3 \times 24 = 72$, these answers check out.

(b) Flavia's room is 8 pedes wide and Octavia's room is 12 pedes wide.

Check: Since $12 - 8 = 4$, Octavia's room is 4 pedes wider than Flavia's room.

Since $12^2 - 8^2 = 144 - 64 = 80$, Octavia's room is 80 pedes quadratum larger in area (square footage) than Flavia's room.

Page LXXX – Age Comparison Problems

(a) Diana is currently VIII (8) years old and Atticus is XXIV (24) years old.

Check:

Now: $8 \times 3 = 24$.

Four years ago: Diana was $8 - 4 = 4$ and Atticus was $24 - 4 = 20$.

Four years ago, $4 \times 5 = 20$.

(b) Priscilla is currently XVII (17) years old and Magnus is VII (7) years old.

Check:

Now: $17 - 7 = 7$.

Five years ago: Priscilla was $17 - 5 = 12$ and Magnus was $7 - 5 = 2$.

Five years ago, $12 = 2 \times 6$.

Page LXXXI – Purchase Problems

(a) XXXVII denarii ($200 - 94 - 69 = 200 - 163 = 37$)

(b) 150 denarii per cow and 75 denarii per goat

Check: $4 \times 150 + 6 \times 75 = 600 + 450 = 1050$

Page LXXXII – Percent Problems

(a) MMCMXXIX denarii ($2900 + 0.1 \times 2900 = 2900 + 29 = 2929$)

(b) MV denarii (the discounted price is $1250 - 0.2 \times 1250 = 1250 - 250 = 1000$, and with tax it comes to $1000 + 0.005 \times 1000 = 1000 + 5 = 1005$)

Page LXXXIII – Fraction Problems

(a) V:IV (there are 36 boys and $81 - 36 = 45$ girls, such that the ratio of girls to boys is $\frac{45}{36}$, which reduces to $\frac{45 \div 9}{36 \div 9} = \frac{5}{4}$, which is 5 to 4, the same as 5:4)

(b) $\frac{I}{II}$ (or $\frac{1}{2}$)

When the daughter eats $\frac{1}{6}$, what remains is $\frac{5}{6}$ before the son eats.

The son eats $\frac{2}{5}$ of $\frac{5}{6}$. Multiply these fractions: $\frac{2}{5} \times \frac{5}{6} = \frac{2}{6} = \frac{1}{3}$ is how much of the pie the son eats.

The daughter eats $\frac{1}{6}$. The son eats $\frac{1}{3}$. Together they eat $\frac{1}{6} + \frac{1}{3} = \frac{1}{6} + \frac{2}{6} = \frac{3}{6} = \frac{1}{2}$.

What remains from the original pie is $1 - \frac{1}{2} = \frac{1}{2}$.

Not yet convinced? Let's look at this problem another way.

Imagine that the mom slices the pie into 6 equal slices.

The daughter eats 1/6 of the pie, which is 1 slice. 5 slices remain.

The son eats 2/5 of these 5 slices. The son eats 2 slices.

The daughter eats 1 slice. The son eats 2 slices. Only 3 of the 6 slices are left.

That's half the pie.

Page LXXXIV – Conversion Problems

(a) MDCCCLXXV pedes ($625 \times 3 = 1875$)

(b) LX librae ($720 \div 12 = 60$)

Page LXXXV – Rate Problems

(a) II horae (note that horae is plural, while hora is singular)

the distance is 12 laps × 20 stadia = 240 stadia

distance = rate × time such that time = $\frac{\text{distance}}{\text{rate}}$

Claudia finishes in $\frac{240}{40} = 6$ horae. Valentina finishes in $\frac{240}{30} = 8$ horae.

Claudia must wait $8 - 6 = 2$ horae for Valentina to finish.

(b) III horae (since every hora, they get $200 + 150 = 350$ stadia closer)

Check: In 3 horae, Marcus will travel $3 \times 150 = 450$ stadia and Julius will travel $3 \times 200 = 600$ stadia, which adds up to $450 + 600 = 1050$ stadia.

Page LXXXVI – Team Problems

(a) IV horae (they can bake $5 + 7 + 8 = 20$ cakes per hora)

Check: In 4 horae, Camilla bakes $5 \times 4 = 20$ cakes, Octavia bakes $7 \times 4 = 28$, and Aurelia bakes $8 \times 4 = 32$ cakes. That's a total of $20 + 28 + 32 = 80$ cakes.

(b) II horae

Lucius build $\frac{1}{3}$ of the wall each hora.

Titus builds $\frac{1}{6}$ of the wall each hora.

In one hora, together they will build $\frac{1}{3} + \frac{1}{6} = \frac{2}{6} + \frac{1}{6} = \frac{3}{6} = \frac{1}{2}$ of the wall.

Therefore, in two hora, the wall will be finished.

ABOUT THE AUTHOR

Dr. Chris McMullen has over 20 years of experience teaching university physics in California, Oklahoma, Pennsylvania, and Louisiana. Dr. McMullen is also an author of math and science workbooks. Whether in the classroom or as a writer, Dr. McMullen loves sharing knowledge and the art of motivating and engaging students.

The author earned his Ph.D. in phenomenological high-energy physics (particle physics) from Oklahoma State University in 2002. Originally from California, Chris McMullen earned his Master's degree from California State University, Northridge, where his thesis was in the field of electron spin resonance.

As a physics teacher, Dr. McMullen observed that many students lack fluency in fundamental math skills. In an effort to help students of all ages and levels master basic math skills, he published a series of math workbooks on arithmetic, fractions, long division, algebra, geometry, trigonometry, and calculus entitled *Improve Your Math Fluency*. Dr. McMullen has also published a variety of science books, including astronomy, chemistry, and physics workbooks.

Author, Chris McMullen, Ph.D.

PUZZLES

The author of this book, Chris McMullen, enjoys solving puzzles. His favorite puzzle is Kakuro (kind of like a cross between crossword puzzles and Sudoku). He once taught a three-week summer course on puzzles. If you enjoy mathematical pattern puzzles, you might appreciate:

300+ Mathematical Pattern Puzzles

Number Pattern Recognition & Reasoning
- Pattern recognition
- Visual discrimination
- Analytical skills
- Logic and reasoning
- Analogies
- Mathematics

ARITHMETIC

For students who could benefit from additional arithmetic practice:
- Addition, subtraction, multiplication, and division facts
- Multi-digit addition and subtraction
- Addition and subtraction applied to clocks
- Multiplication with 10-20
- Multi-digit multiplication
- Long division with remainders
- Fractions
- Mixed fractions
- Decimals
- Fractions, decimals, and percentages
- Grade 6 math workbook

www.improveyourmathfluency.com

MATH

This series of math workbooks is geared toward practicing essential math skills:

- Algebra
- Geometry
- Trigonometry
- Calculus
- Fractions, decimals, and percentages
- Long division
- Multiplication and division
- Addition and subtraction

www.improveyourmathfluency.com

SCIENCE

Dr. McMullen has published a variety of **science** books, including:

- Basic astronomy concepts
- Basic chemistry concepts
- Balancing chemical reactions
- Calculus-based physics textbooks
- Calculus-based physics workbooks
- Calculus-based physics examples
- Trig-based physics workbooks
- Trig-based physics examples
- Creative physics problems
- Modern physics

www.monkeyphysicsblog.wordpress.com

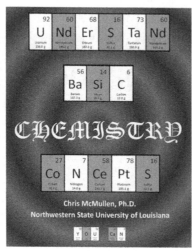

Made in the USA
Monee, IL
30 August 2021